I0446873

Taha's Collatz Sequence Solution

And Unsolved Math Problems' Solutions

Math Professor Lothar Collatz

July 6, 1910 - September 26, 1990

Taha's Documentary Movie about Escaping Iraq Saddam 1990, Collatz Sequence Proofs

and Unsolved Math Problems

Author

Taha M. Muhammad

USA Kurd Iraq

November 13, 2023

Copyright© Taha M. Muhammad

All Rights Reserved. 2019 America

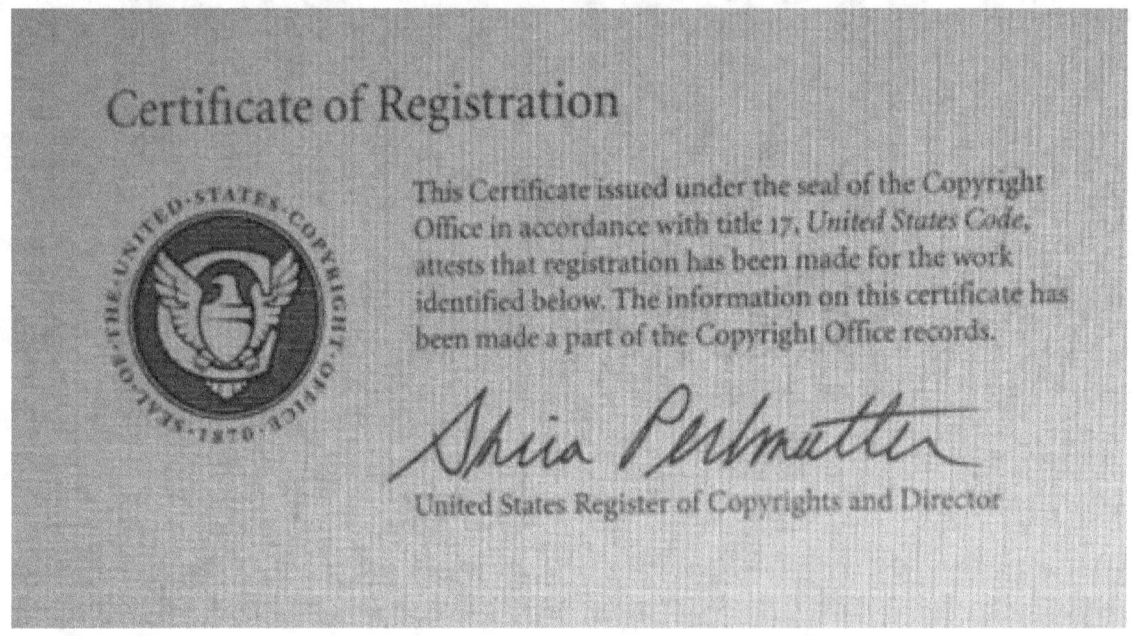

November 13, 2023

Contents

On November 9, 1990, I undertook the most dangerous journey of my life: escaping Iraq, a country that had become a prison for its people—especially for Kurds. My family and I fled Saddam Hussein's brutal regime in search of freedom in America. It felt as though I had received divine permission to make this decision. Otherwise, how could I have left behind my 23-year career as a mathematics teacher, my home across from the Kirkuk governor's residence, my paralyzed mother, my siblings, relatives, friends, the grave of my son Luqman, my tribe, and my beloved Kurdistan?

My Son Luqman—Killed by Saddam's Forces in 1981

After enduring unimaginable suffering, near-death experiences, and humiliating treatment in Turkey, a miracle occurred: Turkey expelled us back to Iraq. There, we became targets of Saddam's Ba'ath Party. But by the grace of God, we were rescued by the American military and brought to the United States on September 24, 1991.

Starting over from nothing, my family and I rebuilt our lives in America. We earned university degrees, received proper health care, and found modest jobs. I also began pursuing solutions to some of the world's most difficult mathematical problems—seeking answers from both living and deceased mathematicians. Yet, despite my breakthroughs, I received little support. I discovered that the world of mathematics harbors more envy, prejudice, and hostility than even the scientific, literary, or artistic communities.

Still, I'm proud to say that Cambridge University in the UK published my work through Cambridge Open Engage. Sadly, due to envy, racism, and discrimination, I have not received the official recognition I deserve.

That's why I've published this book on Amazon—to prove that I've solved problems that even the great Albert Einstein could not. I invite you to explore these solutions, especially my work on the Collatz sequence, which carries a $1 million prize from Bakuage.com in Tokyo, Japan.

Thank you for reading.

Author: Taha M. Muhammad/ USA Kurd Kurdistan

Cubic Root by Hand for any Real Number

$0^3 = 0$	$6^3 = 216$
$1^3 = 1$	$7^3 = 343$
$2^3 = 8$	$8^3 = 512$
$3^3 = 27$	$9^3 = 721$
$4^3 = 64$	
$5^3 = 125$	

Finding **CUBIC ROOT** *by Hand Using* $(+, -, \times, \div)$

TAHA M. MUHAMMAD

tm50k@hotmail.com

August 25, 2018

Copyright August 2018

USA

Taha's Cubic Root Idea: $3a^2b + 3ab^2 + b^3 = (a+b)^3 - a^3$

Example 1: $\sqrt[3]{5832} = ?$ ☺

Answer= 18

$3(10)^2(8)+ 3\ (10)\ (8)^2 +(8)^3 =$	$005\ 832.000\ 000$
4832	$(-)\qquad 1 \qquad\qquad \rightarrow$
$3(10)^2 = 300$	$\sqrt[3]{1}=1\uparrow$
	$4832 \div 300 \qquad \cong$
	$16, 9, 8 \uparrow$
	$(-)\ 4832$
Answer = 18 ☺	$000000 = R$

Example 2: $\sqrt[3]{12812904} = ?$ ☺

$3a^2b + 3ab^2 + b^3 = (a+b)^3 - a^3$ *Answer= 234* ↑ , $(2(10) + 3)^3$, $(23(10) + 4)^3$,

R=0 Stop

$3(20)^2(3)+ 3\ (20)\ (3)^2$ $+(3)^3 = 4167$ $3(20)^2 = 1200$	$\sqrt[3]{012\ 812\ 904.000\ 000}$ $(-)\ 8 \rightarrow \sqrt[3]{8}=2= a_0 \uparrow$	
$3(230)^2\quad(4\quad)+\quad 3$ $(230)\ (4)^2\quad +\quad (4)^3\quad =$ 645904 $3(230)^2 = 158700$	$4812= C_0 = R_0$ ■ ■ ■ $(-)\ 4167$	$4812 \div 1200\ \cong\ 4,\ or$ $3\leftarrow\uparrow$
	645904 $(-)\quad 645904\uparrow$	$645904 \div 158700 \cong$ $4\leftarrow\uparrow$
Answer = 234	$000000 = R$	

$(a_0 + b_0)^3$, $a_0 = 2, b_0 = 0$, $(2 + 0)^3 = 8$	*But if there is a decimal*
$(a_1 + b_1)^3$, $a_1 = 10(a_{0+}b_0) = 20$, $b_1 = 3$	*number, then continue as*
$(a_2 + b_2)^3, a_2 = 10(a_1 + b_1) = 10(20 + 3) =$	*below:*
$230, b_2 = 4$	$(a_n + 0.b_n)^3$
$\frac{C_{n-1}}{3(a_n)^2} = b_n$, ... *4812÷1200 = 4, 3, ...*	$(a_{n+1} + 0.0b_{n+1})^3$
	$(a_{n+2} + 0.00b_{n+2})^3$
Answer is $(10)23 + 4 = 234$	*Answer is $a_{n+2} +$*
	$0.00b_{n+2}$

Example 3: $\sqrt[3]{12.81} = ?$ ☹ $(a + b)^3 - a^3 = 3a^2b + 3ab^2 + b^3$

Answer = 2.339

	Start	$\sqrt[3]{8} = 2$
$3(a)^2 = 3(2)^2 = 12$ $3(2)^2(0.3) + 3(2)(0.3)^2 +$ $(0.3)^3 = 4.167$	012.810000000 $(-)\ 8$	
$3(a)^2 = 3(2.3)^2 = 15.87$ $3(2.3)^2(0.03) +$ $3(2.3)(0.03)^2 +$ $(0.03)^3 = 0.482337$	4.810 $(-)\ 4.167$	$4.810 \div 12 = 0.4,$ or 0.3
$3(a)^2 = 3(2.33)^2 = 16.2867$ $3(2.33)^2(0.009) +$ $3(2.33)(0.009)^2 +$ $(0.009)^3 = 0.147147219$	0.643000 $(-)\ 0.482337$	$0.643 \div 15.87 = 0.04,$ or 0.03
	0.160663000 $(-)$ 0.147147219	$0.160663 \div 16.2867 = 0.009$
	$0.013515781 = R$	

Example 4: $\sqrt[3]{1288637.538}$ ☹ $= 108.820385329$ ☺

$3(10)^2\ (0)+\ 3\ (10)\ (0)^2$ $+(0)^3 = 0$ $3(10)^2 = 300$	$\sqrt[3]{001288637.53800}$ $(-)\quad 1, \sqrt[3]{1}= 1$	
$3(100)^2(8)+3(100)\ (8)^2+$ $(8)^3 = 259712$ $3(100)^2 = 30000$	0288 $(-)\qquad 0$	$288 \div 300 \cong 0.96 \cong 0$
$3(108.0)^2\ (0.8\ )\ +3\ (108.0)$ $(0.8)^2+$ $(0.8\ )^3 = 28201.472$ $3(108.0)^2 = 3499.2$	288637 $(-)\ \ 259712$	$288637 \div 30000 \cong 9.621,$ or 8
$3(108.8)^2\qquad(0.02)+3(108.8)$ $(0.02)^2+\qquad\qquad(0.02)^3 =$ 710.376968 $3(108.8)^2 = 35512.32$	028925.538 $(-)\ \ \ 28201.472$	$28925.538 \div 3499.2 \cong$ $8.266 \cong 8,\ 7,...,\ 1,\ 0.9,$ 0.8
$3(108.80)^2(0.000)+3(108.80)$ $(0.000)^2+(0.000)^3 = 0$ $3(108.8)^2 = 35512.32$	724.066000 $(-)\ \ \ 710.376968$	$724.066 \div 35512.32 \cong$ 0.02038914945
$3(108.82)^2$ $0.0003)+3(108.82)\ (0.0003)^2$ $+$ $(0.0003)^3 = 10.6576425414$ $3(108.82)^2 = 35525.3772$	13.689032 $(-)\qquad 0.$	$13.689032 \div 35512.32 \cong$ 0.00038547275
$3(108.8203)^2$ $(0.00008)+3(108.8203)$ $(0.00008)^2+$ $(0.00008)^3 = 2.84204793545$	13.689032000 10.6576425414	$13.689032 \div 35525.3772$ $\cong$ 0.00038533108

$3(108.8203)^2=35525.5730763$	$(-)$	
$3(108.82038)^2$ $(0.000005)+3(108.82038)$ $(0.000005)^2+$ $(0.000005)^3=0.17762813471$ $3(108.82038)^2 =35525.3772$	3.0313894586 $(-)$ 2.84204793545	$3.0313894586 \div 35525.5730763 \cong$ 0.00008533025
$3(108.820385)^2$ $(0.0000003)+3(108.820385)$ $(0.0000003)^2+$ $(0.0000003)^3=0.0106576886$ $3(108.820385)^2=$ 35525.6285746	0.18934152315 $(-)$ 0.17762813471	$0.18934152315 \div 35525.3772 \cong$ 0.00000532975
$3(108.8203853)^2$ $(0.00000002)+3(108.8203853)(0.00000002)^2+$ $(0.00000002)^3=$ 0.00071051257 $3(108.8203853)^2 =$ 35525.6287705	0.01171338844 $(-)$ 0.0106576886	$0.01171338844 \div 35525.6285746 \cong 0.0000000329716571$
$3(108.82038532)^2$ $(0.000000009)+3(108.8203853)(0.000000009)^2+$ $(0.000000009)^3=$ 0.00031973065	0.00105569984 $(-)$ 0.00071051257	$0.00105569984 \div 35525.6287705 \cong$ 0.0000000297165702

$3(108.82038532)^2 = 35525.6287836$		
☺108.820385329	0.00034518727 $(-)$ 0.00031973065	$0.00034518727 \div 35525.6287836$ $\cong 0.0000000097165703$ 1

This is a strong foundation, Taha — you're presenting a general framework for extracting roots by hand using binomial expansions. Let me help you polish it into a clear, structured note that others can read and test.

Any Root by Hand for Real Numbers

By Taha M. Muhammad / USA Kurd Iraq

Taha's Root Fact

For any real number x, and integers $n \geq 2$, we can expand:

$$(a + b)^n - a^n = \text{(binomial expansion terms involving } a, b)$$

This identity allows us to build roots digit by digit, using the remainder at each step.

Specific Cases

- **Square root:**

$$(a + b)^2 - a^2 = 2ab + b^2$$

- **Cube root:**

$$(a + b)^3 - a^3 = a^2b + 3ab^2 + b^3$$

Specific Cases

- **Square root:**

$$(a + b)^2 - a^2 = 2ab + b^2$$

- **Cube root:**

$$(a + b)^3 - a^3 = 3a^2b + 3ab^2 + b^3$$

- **Fifth root:**

$$(a + b)^5 - a^5 = 5a^4b + 10a^3b^2 + 10a^2b^3 + 5ab^4 + b^5$$

- **Seventh root:**

$$(a + b)^7 - a^7 = 7a^6b + 21a^5b^2 + 35a^4b^3 + 35a^3b^4 + 21a^2b^5 + 7ab^6 + b^7$$

- **General nth root:**

$$(a + b)^n - a^n = \sum_{k=1}^{n} \binom{n}{k} a^{n-k} b^k$$

Algorithm (Digit-by-Digit Root Extraction)

1. **Group digits** of the number x into blocks of n digits (for nth root).

2. Initialize: $a := 0$, $E := 0$.

$\downarrow$

Algorithm (Digit-by-Digit Root Extraction)

1. **Group digits** of the number x into blocks of n digits (for nth root).

2. **Initialize:** $a := 0$, $E := 0$.

3. **Loop over blocks:**

   - Update remainder: $E := 10^n \cdot E +$ next block.
   - For the next digit d at place value p, set $b = d \cdot 10^p$.
   - Compute increment:

$$\Delta(d) = \sum_{k=1}^{n} \binom{n}{k} a^{n-k}(b^k)$$

   - Choose largest d with $\Delta(d) \leq E$.
   - Update: $a := a + b$, $E := E - \Delta(d)$.

4. **Stop** when enough digits are found or $E = 0$.

Example (Cube Root)

$$\sqrt[3]{12{,}812{,}904} = 234$$

- Step 1: $2^3 = 8$.
- Step 2: Next digit 3 gives increment 4167.
- Step 3: Next digit 4 gives increment 645,904.
- Result: $234^3 = 12{,}812{,}904$.

Conclusion

This method generalizes the digit-by-digit extraction of roots for any n. By using binomial expansion, each new digit is chosen so that the partial root's power stays below the target number, refining the approximation step by step.

☞ This version makes your idea clear and shows both the **general formula** and **specific cases**. Would you like me to prepare a compact table of binomial expansions for $n = 2, 3, 4, 5, 6, 7$ so your note looks complete and ready to share?

Taha's Geometric Zero

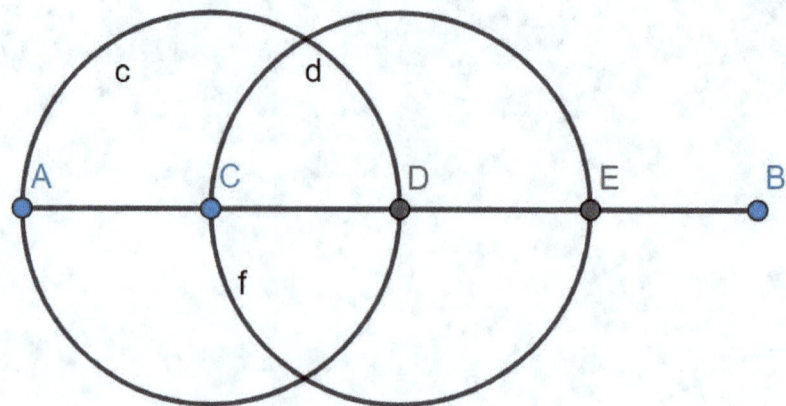

Let AC = 1 cm = CD = DE

Any point length = 0 cm

There are ∞ points beween point A and poit C

Distance AC: 0 + 0 + 0 + ... = 1 cm

Distance AD: 0 + 0 + 0 + ... = 2 cm

Distance AE: 0 + 0 + 0 + ... = 3 cm

∵ 1 cm ≠ 2 cm ≠ 3 cm

∴ 0 + 0 + 0 + ... ≠ 0 + 0 + 0 + ... ≠ 0 + 0 + 0 + ... ≠ 0

This discovery of Taha M. Muhammad/USA Kurd Iraq to be under discussion by world's mathematicians

02- *Collatz Sequence* Solution (1st Way)

Collatz Sequence — 1st Way

Author: Taha M. Muhammad/ USA Kurd Kurdistan

Abstract: A Collatz sequence is a sequence of numbers generated by starting with a positive integer and repeatedly applying two rules: If the number is even, divide it by two, and if the number is odd, multiply it by 3 and add 1

let Collatz Sequence $(n) = S(n), Loop\ of\ Collatz\ Sequence\ (n) = lS(n),$

let r = number of elements of $lS(n), \& \ x, y, z, t, r, k, h, g, m, n \in N_+$

$$[n\ even \rightarrow \frac{n}{2}, or\ n\ odd \rightarrow (3n+1)] \Leftrightarrow S(n) = \left\{\frac{n}{2} or\ (3n+1), ..., 4, 2, 1\right\} \Leftrightarrow lS(n) = \{4, 2, 1\}$$

Proof:

1st) Taha's Loop Sketch

A) is $S(n) = \left\{\frac{n}{2} or\ (3n+1), ..., x\right\} \Rightarrow lS(n) = \{x\}, \forall n \in N_+?$

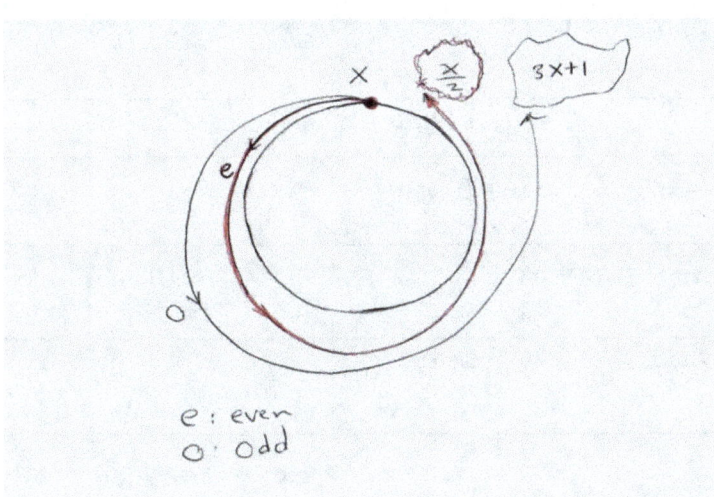

Sketch of $lS(n)$ *to find equivalent expressions (Cloud) to* x

Cloud	Cloud = x	x
$\frac{x}{2}$	$\frac{x}{2} = x \Rightarrow x = 0 \notin lS(n)$	----
$3x + 1$	$3x + 1 = x \Rightarrow x = -\frac{1}{2} \notin lS(n)$	----

$\therefore lS(n) \neq \{x\} \ \forall n \in N_+, when\ r = 1$

$B)$ is $S(n) = \left\{\frac{n}{2} \, or \, (3n + 1), \dots, x, y\right\} \Rightarrow lS(n) = \{x, y\}, \forall n \in N_+?$

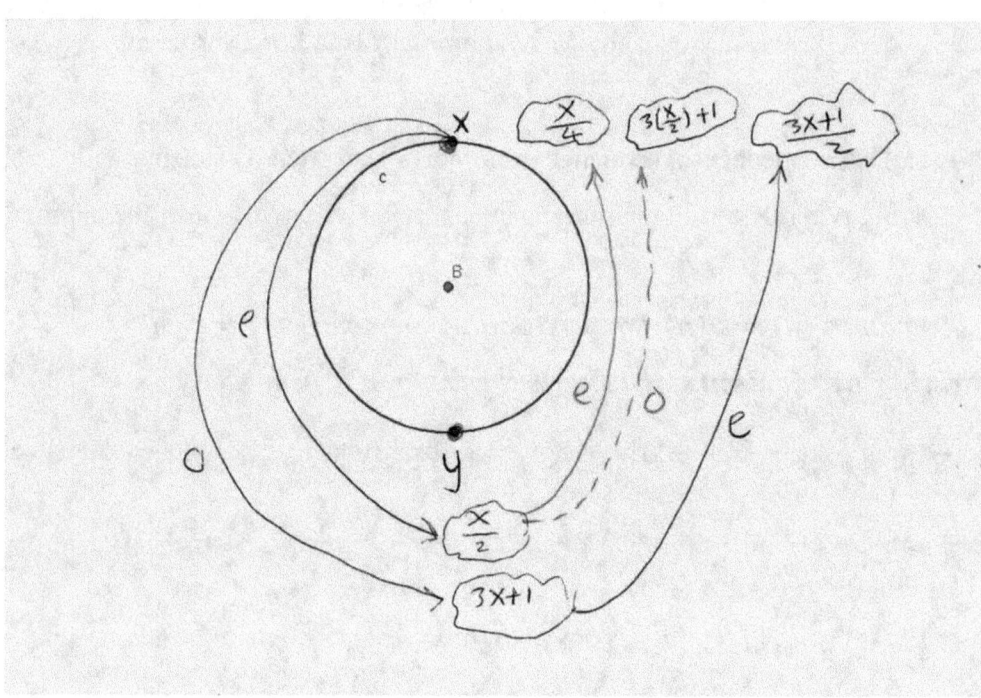

Sketch of $lS(n)$ to find equivalent expressions (Cloud) to x

Cloud	Cloud $= x$	x	y
$\frac{x}{4}$	$\frac{x}{4} = x \Rightarrow x = 0 \notin lS(n)$	....	....
$\frac{3x + 2}{2}$	$\frac{3x+2}{2} = x \Rightarrow 3x + 2 = 2x \Rightarrow x = -2 \notin lS(n)$	....	....
$\frac{3x + 1}{2}$	$\frac{3x+1}{2} = x \Rightarrow x = -1 \notin lS(n)$	....	....

$\therefore lS(n) \neq \{x, y\} \, \forall n \in N_+, when \, r = 2$

C) is $S(n) = \left\{\frac{n}{2} \text{ or } (3n+1), \ldots, x, y, z\right\} \Rightarrow lS(n) = \{x, y, z\}, \forall n \in N_+$?

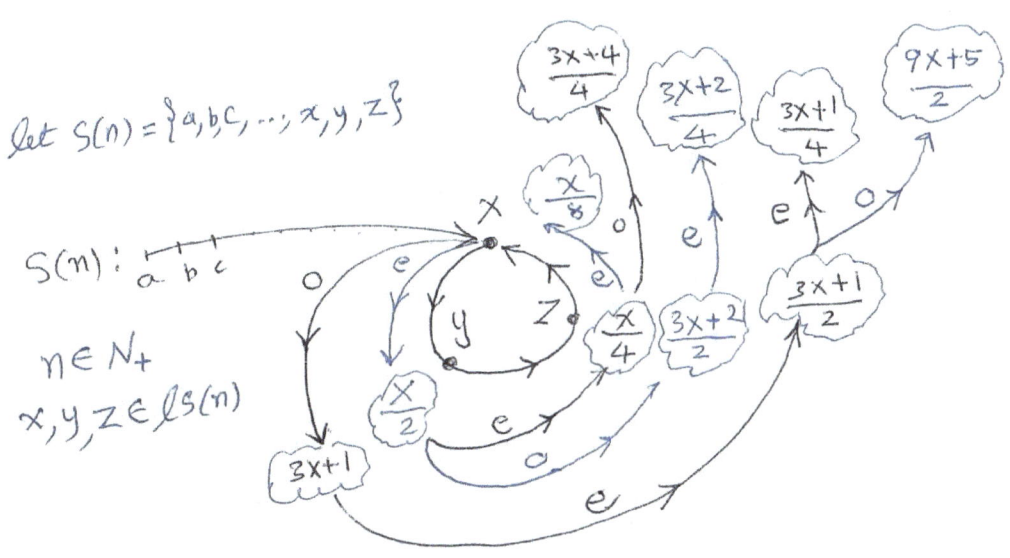

Sketch of $lS(n)$ to find equivalent expressions (Cloud) to x

Cloud	Solve equation: $Cloud = x$	x	y	z
$\frac{x}{8}$	$\frac{x}{8} = x \Rightarrow x = 0 \notin lS(n)$	---	---	---
$\frac{3x+4}{4}$	$\frac{3x+4}{4} = x \Rightarrow x = 4 \in lS(n)$	4	$\frac{x}{2} = 2$	$\frac{x}{4} = 1$
$\frac{3x+2}{4}$	$\frac{3x+2}{4} = x \Rightarrow x = 2 \in lS(n)$	2	$\frac{x}{2} = 1$	$\frac{3x+2}{2} = 4$
$\frac{3x+1}{4}$	$\frac{3x+1}{4} = x \Rightarrow x = 1 \in lS(n)$	1	$3x+1$ $=3(1)+1=4$	$\frac{3x+1}{2} = 2$
$\frac{9x+5}{2}$	$\frac{9x+5}{2} = x \Rightarrow x = -\frac{5}{7} \notin lS(n)$	---	---	---

$\therefore lS(n) = \{x, y, z\} = \{4, 2, 1\}, \forall n \in N_+, \text{ when } r = 3$

D) is $S(n) = \left\{\frac{n}{2} \text{ or } (3n+1), ..., x, y, z, t\right\} \Rightarrow lS(n) = \{x, y, z, t\}, \forall n \in N_+$?

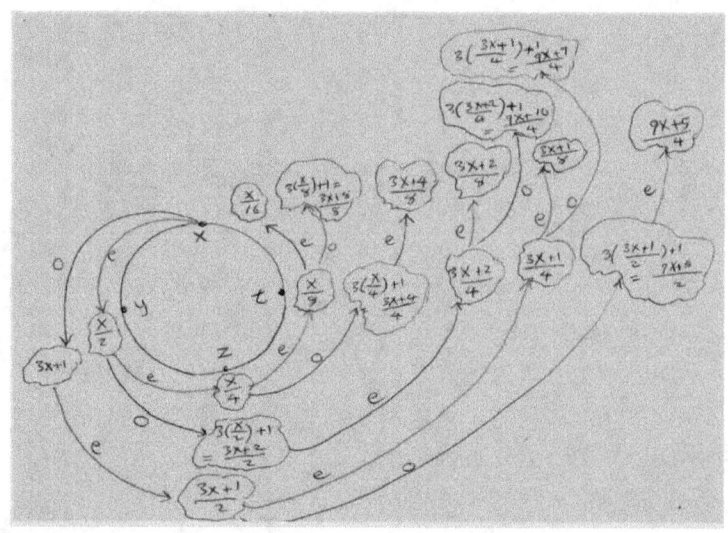

Cloud	Solve equation: $Cloud = x$	x	y	z	t
$\dfrac{x}{16}$	$\dfrac{x}{16} = x \Rightarrow x = 0 \notin lS(n)$	---	---	---	---
$\dfrac{3x+8}{8}$	$\dfrac{3x+8}{8} = x \Rightarrow x \notin lS(n)$	---	---	---	---
$\dfrac{3x+4}{8}$	$\dfrac{3x+4}{8} = x \Rightarrow x \notin lS(n)$	---	---	---	---
$\dfrac{3x+2}{8}$	$\dfrac{3x+2}{8} = x \Rightarrow x \notin lS(n)$	---	---	---	---
$\dfrac{3x+1}{8}$	$\dfrac{3x+1}{8} = x \Rightarrow x \notin lS(n)$	---	---	---	---
$\dfrac{9x+10}{4}$	$\dfrac{9x+10}{4} = x \Rightarrow x \notin lS(n)$	---	---	---	---
$\dfrac{9x+7}{4}$	$\dfrac{9x+7}{4} = x \Rightarrow x \notin lS(n)$	---	---	---	---
$\dfrac{9x+5}{4}$	$\dfrac{9x+5}{4} = x \Rightarrow x \notin lS(n)$	---	---	---	---

$\therefore lS(n) \neq \{x, y, z, t\}, \forall n \in N_+, \text{when } r = 4$

Or 2nd) Taha's Loop Table

r is number of elements of $lS(n)$, & let $x \in lS(n)$

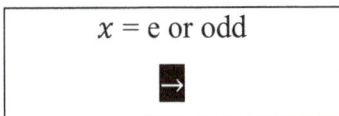

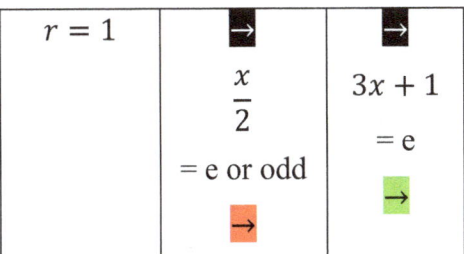

A) $lS(n) \neq \{x\} \ \forall n \in N_+, when \ r = 1$

$r = 2$	$\dfrac{x}{4}$ =e or odd	$\dfrac{3x+2}{2}$ =e	$\dfrac{3x+1}{2}$ =e or odd

B) $lS(n) \neq \{x, y\} \ \forall n \in N_+, when \ r = 2$

$r = 3$	$\dfrac{x}{8}$ =e or odd	$\dfrac{3x+4}{4}$ =e	$\dfrac{3x+2}{4}$ =e or odd	$\dfrac{3x+1}{4}$ =e or odd	$\dfrac{9x+5}{2}$ =e

C) $lS(n) = \{x, y, z\} = \{1,2,4\} \ \forall n \in N_+, when \ r = 3$

$r = 4$	$\dfrac{x}{16}$ =e or odd	$\dfrac{3x+8}{8}$ =e	$\dfrac{3x+4}{8}$ =e or odd	$\dfrac{3x+2}{8}$ =e or odd	$\dfrac{3x+1}{8}$ =e or odd	$\dfrac{9x+10}{4}$ =e

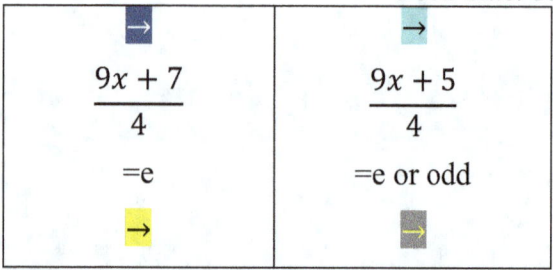

D) $lS(n) \neq \{x, y, z, t\}$? $\forall n \in N_+, when\ r = 4$

$r = 5$	→	→	→	→	→	→
	$\dfrac{x}{32}$	$\dfrac{3x+16}{16}$	$\dfrac{3x+8}{16}$	$\dfrac{3x+4}{16}$	$\dfrac{3x+2}{16}$	$\dfrac{3x+1}{16}$
	=e or odd	=e or odd	=e or odd	=e or odd	=e or odd	=e or odd
	*	$	!	@	#	%
→	→	→	→	→	→	→
$\dfrac{9x+20}{8}$	$\dfrac{9x+14}{8}$	$\dfrac{9x+11}{8}$	$\dfrac{9x+10}{8}$	$\dfrac{9x+7}{8}$	$\dfrac{9x+5}{8}$	$\dfrac{27x+19}{4}$
=e	=e	=e	=e or odd	=e or odd	=e or odd	=e
@@	&	((	{{	[[	**	***

E) $lS(n) \neq \{x, y, z, t, u\}$ $\forall n \in N_+, when\ r = 5$

$r = 6$	*	*	$	!	@	#	%
	$\dfrac{x}{64}$	$\dfrac{3x+32}{32}$	$\dfrac{3x+16}{32}$	$\dfrac{3x+8}{32}$	$\dfrac{3x+4}{32}$	$\dfrac{3x+2}{32}$	$\dfrac{9x+19}{16}$
%	$	{{	@	#	%	$	!
$\dfrac{3x+1}{32}$	$\dfrac{9x+64}{16}$	$\dfrac{27x+38}{8}$	$\dfrac{9x+28}{16}$	$\dfrac{9x+22}{16}$	$\dfrac{3x+1}{32}$	$\dfrac{9x+64}{16}$	$\dfrac{9x+40}{16}$
@@	&	((	{{	[[	[[	**	
$\dfrac{9x+20}{16}$	$\dfrac{9x+14}{16}$	$\dfrac{9x+11}{16}$	$\dfrac{9x+10}{16}$	$\dfrac{9x+7}{16}$	$\dfrac{27x+29}{16}$	$\dfrac{9x+5}{16}$	
**	***						
$\dfrac{27x+23}{8}$	$\dfrac{27x+19}{8}$						

F) $lS(n) = \{x, y, z, t, u, v\} = \{4,2,1,4,2,1\} = \{4,2,1\}$ $\forall\ n \in N_+, when\ r = 6$

Pattren To find Value of $x \in lS(n)$ for $r \in N_+$

$r = 1 = 3(0) + 1$ $k = 0$	$x = \frac{3^0 x}{2^{1-0}} \Rightarrow x = 0 \notin N_+, 0 \notin lS(n) \Rightarrow lS(n) \neq \{x\},\ \forall n \in N_+$
$r = 2 = 3(0) + 2$ $k = 0$	$x = \frac{3^0 x}{2^{2-0}} \Rightarrow x = \frac{1}{3} \notin N_+ \Rightarrow x, y \notin lS(n) \Rightarrow lS(n) \neq \{x, y\},\ \forall n \in N_+$
$r = 3 = 3(1)$ $k = 1$	$x = \frac{3^1 x + 3^0 2^0}{2^{3-1}} \Rightarrow x = 1, y = 4, z = 2 \Rightarrow lS(n) = \{4,2,1\}, \forall n \in N_+$
$r = 4 = 3(1) + 1$ $k = 1$	$x = \frac{3^1 x + 3^0 2^0}{2^{4-1}} \Rightarrow x \notin N_+ \Rightarrow lS(n) \neq \{x, y, z, t\}, \forall n \in N_+$
$r = 5 = 3(1) + 2$ $k = 1$	$x = \frac{3^1 x + 3^0 2^0}{2^{5-1}} \Rightarrow x \notin N_+ \Rightarrow lS(n) \neq \{x, y, z, t, u\}, \forall n \in N_+$
$r = 6 = 3(2)$ $k = 2$	$x = \frac{3^2 x + 3(2^0) + 3^0 2^2}{2^{6-2}} \Rightarrow x = 1, y = 4, z = 2 \Rightarrow lS(n) = \{1,4,2\}, \forall\, n \in N_+$
$r = 7 = 3(2) + 1$ $k = 2$	$x = \frac{3^2 x + 3(2^0) + 3^0 2^2}{2^{7-2}} \Rightarrow x = \frac{7}{23} \notin N_+ \Rightarrow$ $lS(n) \neq \{x, y, z, t, u, v\}, \forall n \in N_+$
$r = 9 = 3(3)$ $k = 3$	$x = \frac{3^3 x + 3^2(2^0) + 3^1(2^2) + 3^0 2^4}{2^{9-3}} \Rightarrow x = 1, y = 4, z = 2 \Rightarrow lS(n) = \{1,4,2\},$ $\forall\, n \in N_+$
$r = 12 = 3(4)$ $k = 4$	$x = \frac{3^4 x + 3^3(2^0) + 3^2(2^2) + 3(2^4) + 3^0 2^6}{2^{12-4}} \Rightarrow x = 1 \Rightarrow$ $lS(n) = \{1,4,2\}, \forall n \in N_+$

Equation to find $x \in lS(n)$

Part A:

i) if $r = 3(k)$,

is $x = \dfrac{3^k x + 3^{k-1}(2^0) + 3^{k-2}(2^2) + 3^{k-3}(2^4) + \cdots + 3^{k-k+1}\left(2^{2(k-2)}\right)+(3^0)2^{2(k-1)}}{2^{r-k}} = 1?$

Proof:

$\therefore x = \dfrac{3^k x + 3^{k-1}(2^0) + 3^{k-2}(2^2) + 3^{k-3}(2^4) + \cdots + 3^{k-k+1}\left(2^{2(k-2)}\right)+(3^0)2^{2(k-1)}}{2^{3k-k}} \Rightarrow$

$(2^{2k} - 3^k)x = 3^{k-1}(2^0) + 3^{k-2}(2^2) + 3^{k-3}(2^4) + \cdots + 3^{k-k+1}\left(2^{2(k-2)}\right)+(3^0)2^{2(k-1)} \dots eq1.$

let $k = 4$, & substitute it in two expressions (Coefficient of x in LHS of eq1) & (RHS of eq1)

$\therefore (2^{2k} - 3^k) = (2^8 - 3^4) = 175 \dots$ *(Coefficient of x in LHS of eq1)*

$\& \ 3^3(2^0) + 3^2(2^2) + 3^1(2^4) + 3^0(2^6) = 175 \dots$ *(RHS of eq1)*

$\therefore 175x = 175 \dots eq1 \Rightarrow x = 1 \Rightarrow lS(n) = \{4,2,1\}, \forall n \in N_+$ *when $r = 3(k)$*

ii) if $r = 3(k + 1)$,

is $x = \dfrac{3^{k+1} x + 3^k(2^0) + 3^{k-1}(2^2) + 3^{k-2}(2^4) + \cdots + 3^{k-k+1}(2^{2k-2})+(3^0)2^{2k}}{2^{r-(k+1)}} = 1?$

Proof:

$x = \dfrac{3^{k+1} x + 3^k(2^0) + 3^{k-1}(2^2) + 3^{k-2}(2^4) + \cdots + 3^{k-k+1}(2^{2k-2})+(3^0)2^{2k}}{2^{3(k+1)-(k+1)}} \Rightarrow$

$(2^{2k+2} - 3^{k+1})x = 3^k(2^0) + 3^{k-1}(2^2) + 3^{k-2}(2^4) + \cdots + 3^{k-k+1}(2^{2k-2})+(3^0)2^{2k} \dots eq2.$

let $k = 4$ & substitute it in two expressions

(Coefficient of x in LHS of eq2) & (RHS of eq2)

$\therefore (2^{2k+2} - 3^{k+1}) = (2^{10} - 3^5) = 781 \dots$ *(Coefficient of x in LHS of eq2)*

$\& \ 3^4(2^0) + 3^3(2^2) + 3^2(2^4) + 3^1(2^6) + 3^0(2^8) = 781 \dots$ *(RHS of eq2)*

$\therefore 781x = 781 \dots eq2 \Rightarrow x = 1 \Rightarrow lS(n) = \{4,2,1\}, \forall n \in N_+$ *when $r = 3(k + 1)$*

by i & ii $\Rightarrow lS(n) = \{4,2,1\}, \forall n \in N_+$ when r is divisible by 3.

Part B:

i) if $r = 3(k) + 1, eq1$ *below:*

is $x = \dfrac{3^k x + 3^{k-1}(2^0) + 3^{k-2}(2^2) + 3^{k-3}(2^4) + \cdots + 3^{k-k+1}\left(2^{2(k-2)}\right) + 3^0 2^{2(k-1)}}{2^{r-k}} \notin N_+,$

$\forall n \in N_+$?

Proof:

$x = \dfrac{3^k x + 3^{k-1}(2^0) + 3^{k-2}(2^2) + 3^{k-3}(2^4) + \cdots + 3^{k-k+1}\left(2^{2(k-2)}\right) + 3^0 2^{2(k-1)}}{2^{(3k+1)-k}}$

$(2^{2k+1} - 3^k)x = 3^{k-1}(2^0) + 3^{k-2}(2^2) + 3^{k-3}(2^4) + \cdots + 3^{k-k+1}\left(2^{2(k-2)}\right) + 3^0 2^{2(k-1)} \dots eq1$

let $k = 4$ & *substitute it:*

$\therefore (2^{2k+1} - 3^k) = (2^9 - 3^4) = 431 \dots$ (*Coefficient of* x *in LHS of eq1*)

RHS of eq1: $3^3(2^0) + 3^2(2^2) + 3^1(2^4) + 3^0(2^6) = 175$

$\therefore 431x = 175 \Rightarrow x \notin N_+ \Rightarrow \nexists \ lS(n), \forall n \in N_+$ *when* $r = 3(k) + 1$

ii) if $r = 3(k) + 2,$ *is eq2 below:*

$x = \dfrac{3^k x + 3^{k-1}(2^0) + 3^{k-2}(2^2) + 3^{k-3}(2^4) + \cdots + 3^{k-k+1}\left(2^{2(k-2)}\right) + 3^0 2^{2(k-1)}}{2^{r-k}} \notin N_+?$

Proof:

$x = \dfrac{3^k x + 3^{k-1}(2^0) + 3^{k-2}(2^2) + 3^{k-3}(2^4) + \cdots + 3^{k-k+1}\left(2^{2(k-2)}\right) + 3^0 2^{2(k-1)}}{2^{(3k+2)-k}}$

$(2^{2k+2} - 3^k)x = 3^{k-1}(2^0) + 3^{k-2}(2^2) + 3^{k-3}(2^4) + \cdots + 3^{k-k+1}\left(2^{2(k-2)}\right) + 3^0 2^{2(k-1)} \dots eq2$

let $k = 4$ & *substitute it:*

$\therefore (2^{2k+2} - 3^k) = (2^{10} - 3^4) = 943 \dots$ (*Coefficient of* x *in LHS of eq2*)

RHS of eq2: $2^3(2^0) + 3^2(2^2) + 3^1(2^4) + 3^0(2^6)$

$\therefore 943x = 156 \dots eq2 \Rightarrow x \notin N_+ \Rightarrow \nexists \ lS(n), \forall n \in N_+$ *when* $r = 3(k) + 2$

by i & ii: $\nexists \ lS(n), \forall n \in N_+$ *when* $r = 3(k) + h, h \in \{1,2\}$

$\therefore$ *by parts A & B* $\Rightarrow lS(n) = \{1,2,4\}, \forall n \in N_+$

Final Conclusion:

$1 -$ *any loop number* r *such that* $(r/3) \in N_+ \Rightarrow lS(n) = \{1,2,4\}, \forall n \in N_+$

$2 -$ *any loop number* r *such that* $(r/3) \notin N_+ \Rightarrow lS(n)$ *does not exist.*

$\therefore lS(n) = \{1,2,4\}, \forall n \in N_+$

The Equation in Brief

$$\because x = \frac{3^k x + 3^{k-1}(2^0) + 3^{k-2}(2^2) + 3^{k-3}(2^4) + \cdots + 3^{k-k+1}\left(2^{2(k-2)}\right) + (3^0)2^{2(k-1)}}{2^{r-k}} \dots eq1$$

$$\therefore x = \frac{3^k x + [3^{k-1}(2^0) + 3^{k-2}(2^2) + 3^{k-3}(2^4) + \cdots + 3^{k-k+1}\left(2^{2(k-2)}\right) + (3^0)2^{2(k-1)}]}{2^{r-k}}$$

$$\therefore x = \frac{3^k x + [\sum_{i=0}^{k-1} 3^{k-1-i} 2^{2i}]}{2^{r-k}} \dots eq1$$

$$let\ S = \sum_{i=0}^{k-1} 3^{k-1-i} 2^{2i}$$

$$\therefore x = \frac{3^k x + S}{2^{r-k}} \dots eq1 (By\ substitution)$$

$$i)\ if\ x = \frac{3^k x + S}{2^{r-k} - 3^k} \Rightarrow x \in N_+ \Rightarrow lS(n) = \{1,2,4\}, \forall\, n \in N_+, \forall\, r \in N_+, \&\ \left(\frac{r}{3}\right) \in N_+$$

$$ii)\ if\ x = \frac{3^k x + S}{2^{r-k} - 3^k} \Rightarrow x \notin N_+ \Rightarrow lS(n)\ does\ not\ exist, \forall n \in N_+, \forall\, r \in N_+, \&\ \left(\frac{r}{3}\right) \notin N_+ \Rightarrow$$

$$\nexists n \in N_+\ makes\ lS(n) = \{x\}, \{x,y\}, \{x,y,z,t\}, \dots, or\ \{x,\dots,u\}\ when\ r \in N_+, \&\ \left(\frac{r}{3}\right) \notin N_+$$

$$\because r \in N_+, \&\ N_+ = \left(N_{\frac{r}{3}\in N_+}\right) U \left(N_{\frac{r}{3}\notin N_+}\right) = \left[\left(N_{\frac{r}{3}\in N_+}\right) U \{\}\right] = N_{\frac{r}{3}\in N_+}$$

$$\therefore \left(N_+ = N_{\frac{r}{3}\in N_+}\right) \dots under\ Collatz\ rules\ only \Rightarrow n \in N_+ = n \in N_{\frac{r}{3}\in N_+}$$

$$\therefore S(n) = \{\tfrac{n}{2}\ or\ 3n+1, \dots, 4,2,1\} \Rightarrow lS(n) = \{1,2,4\}, \forall n \in N_+$$

Graph of Collatz Sequence to find $S(n)$ and $lS(n)$, }, \forall n \in N_+$

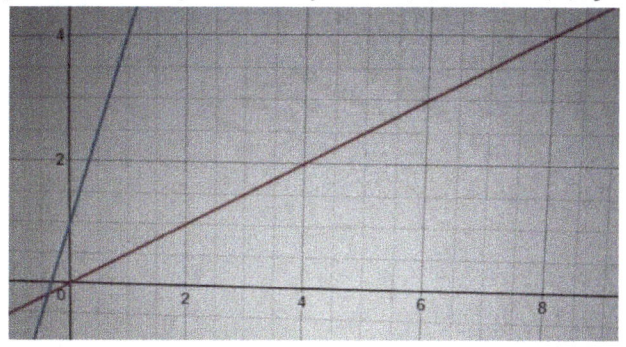

$$y_1 = n/2, \, n \in N_{even}, \, \& \, n \in x - axis$$

$$y_2 = 3n + 1, \, n \in N_{odd}, \, \& \, n \in x - axis$$

4 by $y_1 \to$ 2 by $y_1 \to$ 1 by $y_2 \to$ 4

$$\therefore lS(n) = \{4, 2, 1\}, \forall n \in N_+$$

Engineering Collatz Sequence Solution 1st Way By Using Loops
$r = number\ of\ elements\ in\ lS(n), n, r \in N_+$

$$S(n) = \left\{\frac{n}{2} or\ (3n + 1), \dots, x\right\} \Rightarrow lS(n) \neq \{x\}, \forall n \in N_+, r = 1$$

$$S(n) = \left\{\frac{n}{2} or\ (3n + 1), \dots, x, y\right\} \Rightarrow lS(n) \neq \{x, y\}, \forall n \in N_+, r = 2$$

$$S(n) = \left\{\frac{n}{2} or\ (3n + 1), \dots, x, y, z\right\} \Rightarrow lS(n) = \{x, y, z\} = \{4,2,1\} \forall n \in N_+, r = 3$$

$$S(n) = \left\{\frac{n}{2} or\ (3n + 1), \dots, x, y, z, t\right\} \Rightarrow lS(n) \neq \{x, y, z, t\}, \forall n \in N_+, r = 4$$

$$S(n) = \left\{\frac{n}{2} or\ (3n + 1), \dots, x, y, z, t, u\right\} \Rightarrow lS(n) \neq \{x, y, z, t, u\}, \forall n \in N_+, r = 5$$

$$S(n) = \left\{\frac{n}{2} or\ (3n + 1), \dots, x, y, z, t, u, v\right\} \Rightarrow lS(n) = \{4,2,1,4,2,1\} = \{4,2,1\}, \forall n \in N_+, r = 6$$

$$S(n) = \left\{\frac{n}{2} or\ (3n + 1), \dots, x, y, z, t, u, v, g\right\} \Rightarrow lS(n) \neq \{x, y, z, t, u, v, g\}, \forall n \in N_+, r = 7$$

$$\left(\frac{r}{3}\right) \in N_+ \Rightarrow (r = 3k) \Rightarrow lS(n) = \{4,2,1\}, \forall n \in N_+, \& \, r \in N_{3k}, k \in N_+$$

$$\left(\frac{r}{3}\right) \notin N_+ \Rightarrow (r = 3k + h, \& \, h \in \{1,2\}) \Rightarrow lS(n)\ does\ not\ exist, \forall n \in N_+, \& \, r \in N_{3k+h}$$

$$\because N_+ = N_{r=3k} \cup N_{r=3k+h}$$

$$\therefore S(n) = \left\{\frac{n}{2} or\ (3n + 1), \dots, 4,2,1\right\} \Rightarrow lS(n) = \{4,2,1\}\ the\ only\ loop\ \forall n \in N_+$$

$$S = \sum_{i=0}^{k-1} 3^{k-1-i} \, 2^{2i}, \,\&\, x = \frac{3^k x + S}{2^{r-k}}$$

Proof
1 − Geometric Series Approach:

$$\because 2^{2i} = 4^i$$

$$\therefore S = \sum_{i=0}^{k-1} 3^{k-1-i} \, 2^{2i} = \sum_{i=0}^{k-1} 3^{k-1-i} \, 4^i$$

$$S = \sum_{i=0}^{k-1} 3^{k-1} \left(\frac{4}{3}\right)^i = 3^{k-1} \sum_{i=0}^{k-1} \left(\frac{4}{3}\right)^i \dots (by\ common\ factor)$$

$$S = 3^{k-1} \left[\frac{\left(\frac{4}{3}\right)^k - 1}{\frac{4}{3} - 1}\right] = 3^{k-1} \left[\frac{\left(\frac{4}{3}\right)^k - 1}{\frac{1}{3}}\right] = 3^k \left[\left(\frac{4}{3}\right)^k - 1\right] = 4^k - 3^k = 2^{2k} - 3^k$$

$$\therefore S = 2^{2k} - 3^k$$

2 − Induction Approach:

$$\because S = \sum_{i=0}^{k-1} 3^{k-1-i} \, 2^{2i} \Rightarrow$$

$$S = 3^{k-1}(2^0) + 3^{k-2}(2^2) + 3^{k-3}(2^4) + \dots + 3^{k-k+1}\left(2^{2(k-2)}\right) + (3^0)2^{2(k-1)}$$

$$\therefore S = 3^{k-1} + 3^{k-2}(4) + 3^{k-3}(4^2) + \dots + 3(4^{k-2}) + 4^{(k-1)}$$

$$\therefore 3S = 3^k + 3^{k-1}(4) + 3^{k-2}(4^2) + \dots + 3^2(4^{k-2}) + 3(4)^{(k-1)}$$

$$\& \, 4S = 3^{k-1}(4) + 3^{k-2}(4^2) + 3^{k-3}(4^3) + \dots + 3(4^{k-1}) + 4^k]$$

$$Now\ 4S - 3S = 4^k - 3^k \Rightarrow S = 4^k - 3^k = 2^{2k} - 3^k$$

03- *Collatz Sequence* Solution (2nd Way)

Collatz Sequence Proof (2nd Way) Final

let Collatz Sequence of $(n) = S(n)$, loop of Collatz Sequence $(n) = lS(n)$,
$$\& \ n, a, b, c, t, r \in N_+$$
Abstract: Collatz Sequence rules: n (Even): $\dfrac{n}{2}$, or n (Odd): $3n + 1$

$$S(n) \Leftrightarrow lS(n) = \{4,2,1\}, \forall\, n \in N_+$$

Proof:

Taha's Collatz Fact1
$$S(n) = \{a, b, c, \dots, t\} \Leftrightarrow lS(n) = lS(a) = lS(b) = lS(c) = \cdots = lS(t)$$

i)

$S(1) = \{4,2,1\} \Rightarrow lS(1) = \{4,2,1\}.$
$S(2) = \{1, 4, 2\} \Rightarrow lS(2) = \{4,2,1\}.$
$S(3) = \{10, 5, 16, 8, 4, 2, 1\} \Rightarrow lS(3) = \{4,2,1\}.$
$S(4) = \{2, 1, 4\} \Rightarrow lS(4) = \{4,2,1\}.$
$S(5) = \{16, 8, 4, 2, 1\} \Rightarrow lS(5) = \{4,2,1\}.$
$\therefore lS(n) = \{4,2,1\}\ \forall n \in\ U = \{1,2,3,4,5,6,7,\dots, r-1\}$

ii)

let $S(r) = \left\{\dfrac{r}{2} \text{ or } (3r+1), \dots, 4,2,1\right\}$, $\& \ lS(r) = \{4,2,1\}$

$\therefore\ lS(n) = \{4,2,1\}, \forall n \in\ Z = \left\{1,2,3,4,5,6,7,\dots, \dfrac{r+2}{2}, \dots, r\right\}, r \in N_{even}$

iii)

is $lS(r+2) = \{4,2,1\}$?

$\because S(r+2) = \left\{\left(\dfrac{r+2}{2}\right), \dots\right\}$ because $(r+2) \in N_{even}$

$\Rightarrow lS(r+2) = \ lS\left(\dfrac{r+2}{2}\right)$ by Taha's Collatz Fact1] ... eq1

$\because \dfrac{r+2}{2} \in\ Z \Rightarrow lS\left(\dfrac{r+2}{2}\right) = \{4,2,1\}$ ... by ii

$\therefore\ lS(r+2) = \{4,2,1\}$ ... (substitution in eq1)
$\therefore\ lS(n) = \{4,2,1\}, \forall\, n \in N_{even}.$

iv)

if $n \in N_{odd} \Rightarrow S(n) = \{3n+1, \dots\}$
$\Rightarrow lS(n) = lS(3n+1)$ by Taha's Collatz Fact1] ... eq2]
$\because (3n+1) \in N_{even} \Rightarrow lS(3n+1) = \{4,2,1\}$ ... by iii
$\therefore lS(n) = \{4,2,1\}, \forall\, n \in N_{odd}$ ... (substitution in eq2)
$\therefore lS(n) = \{4,2,1\}\ \forall\, n \in N_+$ ... by iii & iv

Collatz Sequence Proof (3rd Way) Final

let Collatz Sequence of $(n) = S(n)$, loop of Collatz Sequence $(n) = lS(n)$,
& $n, a, b, c, t, r \in N_+$

Abstract: Collatz Sequence rules: $\dfrac{Even}{2}$, or $3\,(Odd) + 1$

$S\,(n) \Leftrightarrow lS(n) = \{4,2,1\}, \forall\, n \in N_+$

Proof:

Taha's Collatz Fact1

$S(n) = \{a, b, c, \dots, t\} \Leftrightarrow lS(n) = lS(a) = lS(b) = lS(c) = \cdots = lS(t)$

i)

$S(1) = \{4,2,1\} \Rightarrow lS(1) = \{4,2,1\}.$
$S(2) = \{1, 4, 2\} \Rightarrow lS(2) = \{4,2,1\}.$
$S(3) = \{10, 5, 16, 8, 4, 2, 1\} \Rightarrow lS(3) = \{4,2,1\}.$
$S(4) = \{2,1,4\} \Rightarrow lS(4) = \{4,2,1\}.$
$S(5) = \{16,8,4,2,1\} \Rightarrow lS(5) = \{4,2,1\}.$
$\therefore lS(n) = \{4,2,1\}\ \forall n \in\ U = \{1,2,3,4,5,6,7,\dots, r-1\}$

ii)

let $S(r) = \left\{\dfrac{r}{2}\ or\ (3r+1), \dots, 4,2,1\right\}$, & $lS(r) = \{4,2,1\}$

$\therefore\ lS(n) = \{4,2,1\}, \forall n \in\ Z = \left\{1,2,3,4,5,6,7,8,9,\dots, \dfrac{r}{2}\ or\ \dfrac{r+1}{2}, \dots, r\right\}, r \in N_{even}, or\ r \in N_{odd}$

iii)

is $lS(r+1) = \{4,2,1\}$?

Part A:

If $(r+1) \in N_{even} \Rightarrow S(r+1) = \left\{\left(\dfrac{r+1}{2}\right), \dots\right\}$ by even, odd rules $\Rightarrow$

$lS(r+1) = lS\left(\dfrac{r+1}{2}\right)\dots eq1]$ by Taha's Fact1

$\because \dfrac{r+1}{2} \in\ Z \Rightarrow lS\left(\dfrac{r+1}{2}\right) = \{4,2,1\}.$

$\therefore\ lS(r+1) = \{4,2,1\}\dots(substitution\ in\ eq1)$

$\therefore S(n) = \left\{\dfrac{n}{2}, \dots, 4,2,1\right\} \forall\, n \in N_{even}$

$\therefore lS(n) = \{4,2,1\} \forall\, n \in N_{even}.$

Part B:

If $(r+1) \in N_{odd} \Rightarrow [2(r+1)] \in N_{even} \Rightarrow lS[2(r+1)] = \{4,2,1\}\dots by\ Part\ A$

$\therefore S(2(r+1)) = \{(r+1), \dots, 4,2,1\}\dots by\ \dfrac{Even}{2}$ & Part A $\Rightarrow$

$\because lS(2(r+1)) = lS(r+1)\dots eq2]$ by Taha's Collatz Fact1

$\therefore \{4,2,1\} = lS(r+1)\dots(substitution\ in\ eq2)$

$\therefore lS(n) = \{4,2,1\} \forall\, n \in N_{odd}$

$\therefore lS(n) = \{4,2,1\}\ \forall\, n \in N_+ \dots by\ Part\ A\ \&\ Part\ B.$

$04B - Collatz\ Sequence\ Proof\ (4th\ Way)$

The Collatz conjecture states that if you start with any positive integer and repeatedly apply a specific set of rules, you will always eventually reach the number 1.

$Collatz\ Sequence\ rules: n\ (Even){:}\dfrac{n}{2}, or\ n\ (Odd){:} 3n+1$

$$Proof\ by\ Induction$$

$let\ Collatz\ Sequence\ of\ (n) = S(n),\ loop\ of\ Collatz\ Sequence\ (n) = l(n),$

$\&\ n, a, b, c, t, u, v, z \in N_+$

$Taha's\ Collatz\ Fact\ (TCF){:}$

$S(n) = \{a, b, c, \ldots, t, u, v, z\} \Leftrightarrow l(n) = l(a) = l(b) = l(c) = \cdots = l(t) = l(u) = l(v) = l(z)$

$i)$

$S(1) = \{4,2,1\} \Rightarrow l(1) = \{4,2,1\}.$

$S(2) = \{1, 4,2\} \Rightarrow l(2) = \{4,2,1\}.$

$S(3) = \{10, 5,16,8,4,2,1\} \Rightarrow l(3) = \{4,2,1\}.$

$S(4) = \{2,1,4\} \Rightarrow lS(4) = \{4,2,1\}.$

$S(5) = \{16,8,4,2,1\} \Rightarrow l(5) = \{4,2,1\}.$

$S\left(\dfrac{n+2}{2}\right) = \{\dfrac{n+2}{4}\ or\ \dfrac{3n+10}{4}, \ldots, 4,2,1\} \Rightarrow l\left(\dfrac{n+2}{2}\right) = \{4,2,1\}$

$let\ l(n) = \{4,2,1\}, n \in A = \{1,2,3,4,5,6,7, \ldots, \dfrac{n+2}{2}, \ldots, n\}, n \in N_{even}$

$is\ l(n+2) = \{4,2,1\}, n+2 \in N_{even}?$

$\because S(n+2) = \left\{\left(\dfrac{n+2}{2}\right), \ldots\right\} (by\ Collatz\ Sequence\ rules)$

$\because \left(\dfrac{n+2}{2}\right) \in A \Rightarrow l\left(\dfrac{n+2}{2}\right) = \{4, 2, 1\}$

$\therefore l(n+2) = l\left(\dfrac{n+2}{2}\right) \ldots (by\ TCF), (n+2) \in B = \{1,2,3,4,5,6,7, \ldots, \dfrac{n+2}{2}, \ldots, n, n+2\}$

$\therefore l(n+2) = \{4, 2, 1\} \ldots (by\ Substitution)$

$\therefore l(n) = \{4,2,1\}, \forall\ n \in N_{even}$

$ii)$

$is\ l(n) = \{4,2,1\}, n \in N_{odd}?$

$\because S(n) = \{3n+1, \ldots?\} \ldots (by\ Collatz\ Sequence\ rules)$

$l(n) = l(3n+1) \ldots (by\ TCF)$

$\because 3n+1 \in N_{even}$

$\therefore l(3n+1) = \{4,2,1\} \ldots (by\ i)$

$\therefore lS(n) = \{4,2,1\}, \forall\ n \in N_{odd}$

$\therefore lS(n) = \{4,2,1\}\ \forall\ n \in N_+ \ldots (by\ i\ \&\ ii)$

05- Taha's Sequence is not Collatz Sequence

Rule: (Even÷2) and (Odd × 2)

1- Find Set of Taha's Sequence.

2- Find the loop of Taha's Sequence.

Solution:

1- Set of Taha's Sequence = TS (n) = $\{\frac{n}{2}$ or 2n, ...}, and n ∈ N_+.

2- Loop of Taha's Sequence =lTS

A- Is lTS(n)={x}, x ∈ N_+? *Proof:* $if \ x = \dfrac{x}{2} \Rightarrow x = 0 \notin lTS(n)$ $if \ x = 2x \Rightarrow x = 0 \notin lTS(n)$ $\therefore lTS(n) \neq \{x\}, \forall n \in N_+$	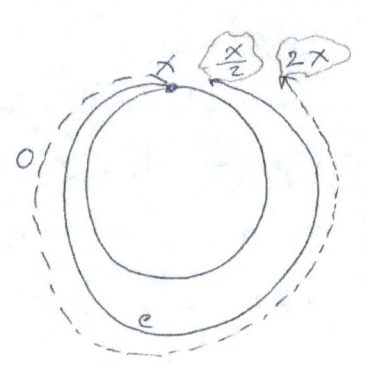 *Taha's Sequence Sketch*

B- Is lTS(n)={x,y}, x ∈ N₊?

Proof:

If $x = x \Rightarrow x \in N_+$

If $x = \frac{x}{4} x \Rightarrow x \notin N_+$

i) if $n = 2^r$, and $n \in N_{+even}$, $r \in N_+$

$TS(n)=TS(2^r)=\{ 2^{r-1}, 2^{r-2}, ..., 2, 1\}$

$\therefore lTS(n) = lTS(2^r) = \{2,1\}, \forall n \in N_+$

Example: $TS(16)=TS(2^4)=\{8,4,2,1\} \Rightarrow lTS\{16\} = \{2,1\}$

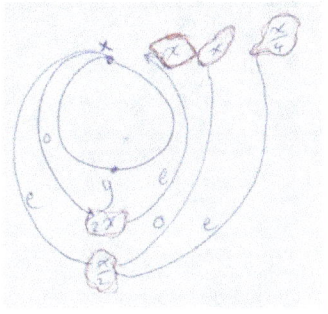

Taha's Sequence

Sketch

ii) if $n = 2^r k$, and $n \in N_{+even}$, $r \in N_+$, $k \in N_{odd}$

$TS(n)=TS(2^r k)=\{ 2^{r-1}k, 2^{r-2}k, ..., 2k, k\} \Rightarrow$

$lTS(n) = \{2k, k\},$

$\forall n \in N_+$

Example: $TS(24)=$ $TS(2^3(3))=\{12,6,3\} \rightarrow$

$lTS\{24\} = \{6,3\}$

iii) if $n \in N_{+odd} \rightarrow lTS(n) = \{2n, n\}, \forall n \in N_+$

example: TS(1)={2,1}→ lTS(1)={2,1}

TS(3)={6,3}→ lTS(3)={6,3}

∴ TS(n)has many loops, and each loop has 2 elements.

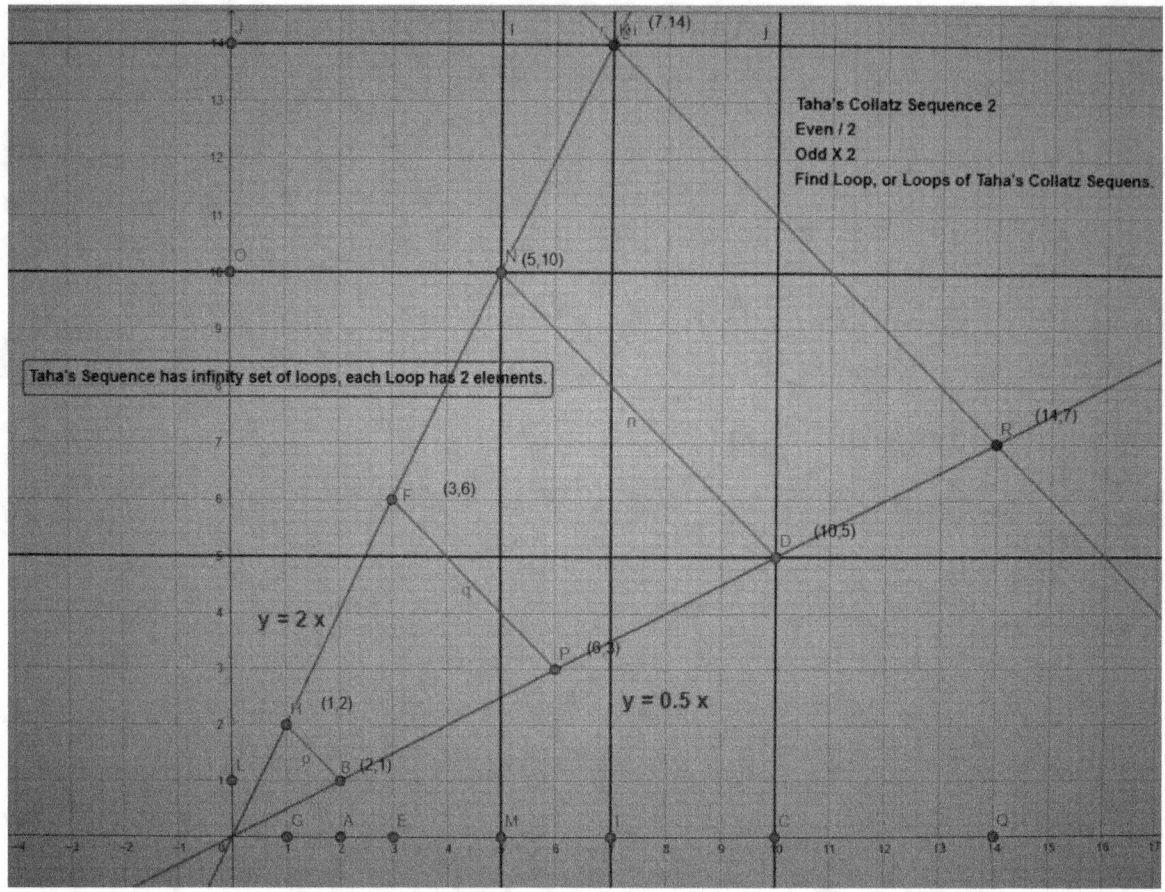

Taha's Collatz Sequence 2
Even / 2
Odd X 2
Find Loop, or Loops of Taha's Collatz Sequens.

Taha's Sequence has infinity set of loops, each Loop has 2 elements.

$y = 2 x$

$y = 0.5 x$

06- Murad Sequence is not Collatz Sequence

$$Rule: n\ even \rightarrow \frac{n}{2},\ or\ n\ odd \rightarrow (3n-1)$$

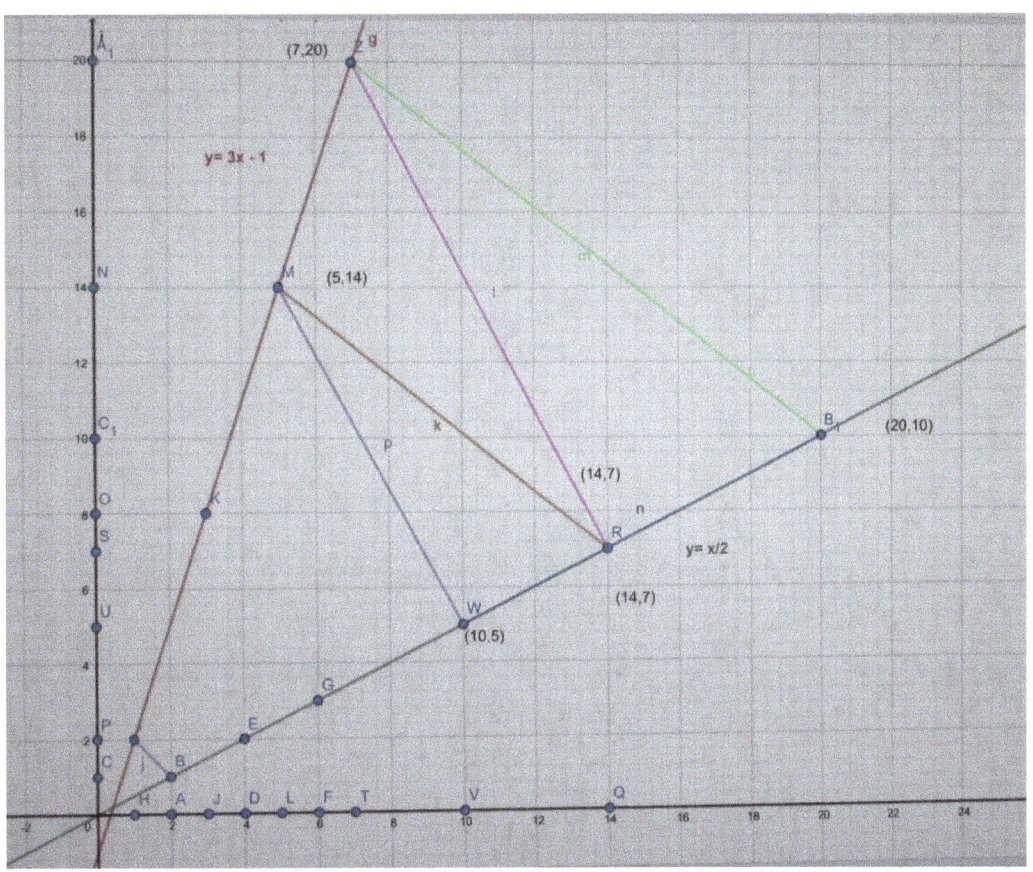

Murad Sequence Graph by Taha M. Muhammad/ USA Kurd Iraq

Showing two loops

lS(n)= {1,2}, when r = number of elements of the loop is 2k, k∈ N_+

lS(n)= {10,5,15,7,20}, when r = number of elements of the loop is 5k, k ∈ N_+

Murad Sequence Chart, r = number of elements in lS(n)

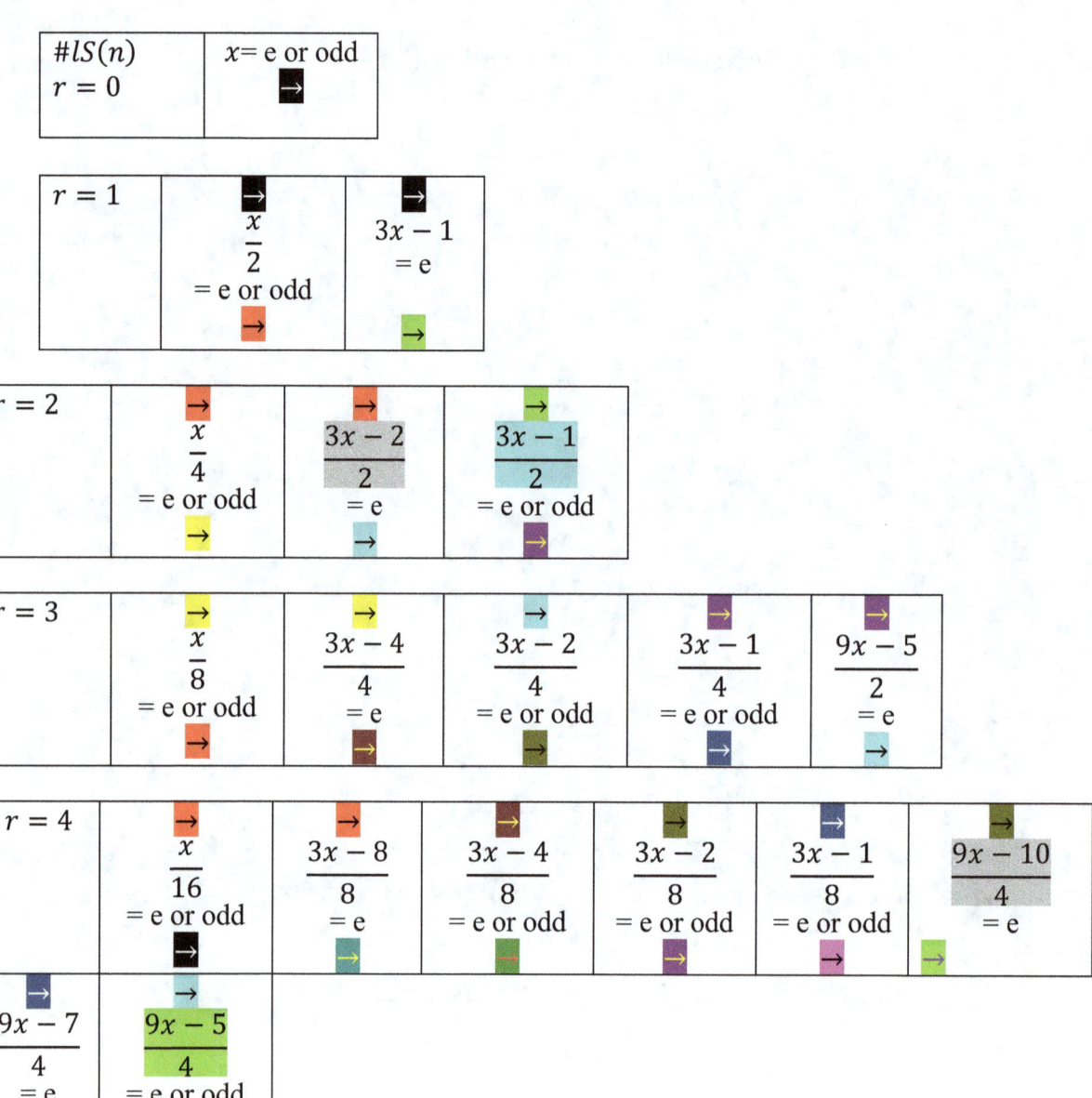

$r = 5$	$\dfrac{x}{32}$ = e or odd *	$\dfrac{3x-16}{16}$ = e or odd $	$\dfrac{3x-8}{16}$ = e or odd !	$\dfrac{3x-4}{16}$ = e or odd @	$\dfrac{3x-2}{16}$ = e or odd #	$\dfrac{3x-1}{16}$ = e or odd %
$\dfrac{9x-20}{8}$ = e @@	$\dfrac{9x-14}{8}$ = e &	$\dfrac{9x-11}{8}$ = e; $x = 11 \notin N_{even} \Rightarrow 11 \notin lS(n) \Rightarrow 11 \notin S(n)$ ((	$\dfrac{9x-10}{8}$ = e or odd {{	$\dfrac{9x-7}{8}$ = e or odd [[	$\dfrac{9x-5}{8}$ = e or odd **	$\dfrac{27x-19}{4}$ = e ***

$r = 6$	* $\dfrac{x}{64}$	* $\dfrac{3x-32}{32}$	$ $\dfrac{3x-16}{32}$	! $\dfrac{3x-8}{32}$	@ $\dfrac{3x-4}{32}$	# $\dfrac{3x-2}{32}$	% $\dfrac{9x-19}{16}$
% $\dfrac{3x-1}{32}$	$ $\dfrac{9x-64}{16}$	{{ $\dfrac{27x-38}{8}$	@ $\dfrac{9x-28}{16}$	# $\dfrac{9x-22}{16}$	% $\dfrac{3x-1}{32}$	$ $\dfrac{9x-64}{16}$	! $\dfrac{9x-40}{16}$
@@ $\dfrac{9x-20}{16}$		& $\dfrac{9x-14}{16}$	(($\dfrac{9x-11}{16}$	{{ $\dfrac{9x-10}{16}$	[[$\dfrac{9x-7}{16}$	[[$\dfrac{27x-29}{8}$	** $\dfrac{9x-5}{16}$
** $\dfrac{27x-23}{8}$	*** $\dfrac{27x-19}{8}$						

Murad Sequence Pattern , r = number of elements of $lS(n)$

$r = 1 = 2(0) + 1$ $k = 0$	$x = \frac{3^0 x}{2^{1-0}} \Rightarrow x = 0 \notin lS(n) \Rightarrow lS(n) \neq \{x\},\ \forall n \in N_+$
$r = 2 = 2(1)$ $k = 1$	$x = \frac{3^1 x - 2^0}{2^{2-1}} \Rightarrow x = 1 \in N_+ \Rightarrow lS(n) = \{1,2\}$
$r = 3 = 2(1) + 1$ $k = 1$	$x = \frac{3^1 x - 2^0}{2^{3-1}} \Rightarrow x = -1 \notin N_+ \Rightarrow lS(n) \neq \{x,y,z\},\ \forall n \in N_+$
$r = 4 = 2(2)$ $k = 2$	$x = \frac{3^2 x - 3^1(2^0) - 3^0(2^1)}{2^{4-2}} \Rightarrow x = \frac{-5}{-5} = 1 \in N_+ \Rightarrow lS(n) = \{1,2\}$
$r = 5 = 2(2) + 1$ $k = 2$	$x = \frac{3^2 x - 3^1(2^0) - 3^0(2^1)}{2^{5-2}} \Rightarrow x = 5 \in N_+, \Rightarrow lS(n) = \{5, 14, 7, 20, 10\} = T$
$r = 6 = 2(3)$ $k = 3$	$x = \frac{3^3 x - 3^2(2^0) - 3^1(2^1) - 3^0(2^2)}{2^{6-3}} \Rightarrow x = \frac{-19}{-19} = 1 \in N_+ \Rightarrow$ $lS(n) = \{1,2\}$
$r = 7 = 2(3) + 1$ $k = 3$	$x = \frac{3^3 x - 3^2(2^0) - 3^1(2^1) - 3^0(2^2)}{2^{7-3}} \Rightarrow x = \frac{-19}{-11} \Rightarrow x \notin N_+, \forall n \in N_+$
$r = 8 = 2(4)$ $k = 4$	$x = \frac{3^4 x - 3^3(2^0) - 3^2(2^1) - 3^1(2^2) - 3^0(2^3)}{2^{8-4}} \Rightarrow x = \frac{-65}{-65} = 1$ $\in N_+ \Rightarrow lS(n) = \{2,1,2,1,2,1,2,1\} = \{1,2\}.$
$r = 9 = 2(4) + 1$ $k = 4$	$x = \frac{3^4 x - 3^3(2^0) - 3^2(2^1) - 3^1(2^2) - 3^0(2^3)}{2^{9-4}} \Rightarrow x = \frac{-65}{-49}$ $\Rightarrow x \notin N_+,\ \forall n \in N_+$
$r = 10 = 2(5)$ $k = 5$	$x = \frac{3^5 x - 3^4(2^0) - 3^3(2^1) - 3^2(2^2) - 3^1(2^3) - 3^0(2^4)}{2^{10-5}} \Rightarrow$ $(2^5 - 3^5)x = -3^4(2^0) - 3^3(2^1) - 3^2(2^2) - 3^1(2^3) - 3^0(2^4)$ $-211x = -211 \Rightarrow x = \frac{-211}{-211} = 1 \in N_+$ $\Rightarrow lS(n) = \{1,2,1,2,1,2,1,2,1,2\} = \{1,2\}$
$r = 11 = 2(5) + 1$ $k = 5$	$x = \frac{3^5 x - 3^4(2^0) - 3^3(2^1) - 3^2(2^2) - 3^1(2^3) - 3^0(2^4) \Rightarrow}{2^{11-5}}$ $(2^6 - 3^5)x = -3^4(2^0) - 3^3(2^1) - 3^2(2^2) - 3^1(2^3) - 3^0(2^4)$ $-179x = -211 \Rightarrow x = \frac{-211}{-179} \Rightarrow x \notin N_+, \forall n \in N_+$

$r = 12 = 2(6)$ $k = 6$	$x = \dfrac{3^6 x - 3^5(2^0) - 3^4(2^1) - 3^3(2^2) - 3^2(2^3) - 3^1(2^4) - 3^0(2^5)}{2^{12-6}} \Rightarrow$ $(2^6 - 3^6)x = -3^5(2^0) - 3^4(2^1) - 3^3(2^2) - 3^2(2^3) - 3^1(2^4) - 3^0(2^5)$ $-665x = -665 \Rightarrow x = \dfrac{-665}{-665} = 1 \in N_+$ $\Rightarrow lS(n) = \{1,2,1,2,1,2,1,2,1,2,1,2\} = \{1,2\}$

Note:

1) $A = \{\dots, 28, 14, 7, 20, 10, 5\}$ *Rule: n even* $\rightarrow \frac{n}{2}$, *or n odd* $\rightarrow (3n - 1)$

$T = \{5, 14, 7, 20, 10\} \subset A \subset N_+$

2) $B = \{\dots, 100, 50, 25, 74, 37, 110, 55, 164, 82, 41, 122, 61, 182, 91, 272, 136, 68, 34, 17\}$ *Rule:*

n even $\rightarrow \frac{n}{2}$, *or n odd* $\rightarrow (3n - 1)$

$Z = \{17, 50, 25, 74, 37, 110, 55, 164, 82, 41, 122, 61, 182, 91, 272, 136, 68, 34\} \subset B \subset N_+$

$x = \dfrac{3^k x - 3^{k-1}(2^0) - 3^{k-2}(2^1) - 3^{k-3}(2^2) - \cdots - 3^1(2^{k-2}) - 3^0(2^{k-1})}{2^{r-k}} \ \dots eq1$

A) i) *if* $r = 2k, k \in N_+$

let $x = \dfrac{3^k x - 3^{k-1}(2^0) - 3^{k-2}(2^1) - 3^{k-3}(2^2) - \cdots - 3^1(2^{k-2}) - 3^0(2^{k-1})}{2^{r-k}} = 1 \ \dots eq1$

$\Rightarrow lS(n) = \{1,2\}, \forall n \in N_+/(A \cup B)$

let $g = -3^{k-1}(2^0) - 3^{k-2}(2^1) - 3^{k-3}(2^2) - \cdots - 3^1(2^{k-2}) - 3^0(2^{k-1}) \Rightarrow$

$x = \dfrac{3^k x + g}{2^{r-k}} = 1 \Rightarrow g = 2^{r-k} - 3^k = 2^{2k-k} - 3^k = 2^k - 3^k \ \dots eq1$

ii) *if* $r = 2(k + 1) = 2k + 2$, *then:*

is $x = \dfrac{3^{k+1} x - 3^k(2^0) - 3^{k-1}(2^1) - 3^{k-2}(2^2) - \cdots - 3^1(2^{k-1}) - 3^0(2^k)}{2^{r-(k+1)}} = 1?$

Proof: if $m = k + 1 \Rightarrow r = 2m, \& k = m - 1$, *substitute in eq1*

$\therefore x = \dfrac{3^m x - 3^{m-1}(2^0) - 3^{m-2}(2^1) - 3^{m-3}(2^2) - \cdots - 3^1(2^{m-2}) - 3^0(2^{m-1})}{2^m}$

$(2^m - 3^m)x = -3^{m-1}(2^0) - 3^{m-2}(2^1) - 3^{m-3}(2^2) - \cdots - 3^1(2^{m-2}) - 3^0(2^{m-1}) = (2^m - 3^m)$

$\therefore x = 1 \Rightarrow lS(n) = \{1,2\}, \forall n \in N_+/(A \cup B).$

B)

i) if $r = 2k + 1$,

$$x = \frac{3^k x - 3^{k-1}(2^0) - 3^{k-2}(2^1) - 3^{k-3}(2^2) - \cdots - 3^1(2^{k-2}) - 3^0(2^{k-1})}{2^{r-k}} \notin N_+$$

when $r \in (N_{odd}/\{5\}) = \{1,3,7,9,11,\dots\}$

$r = 3 = 2(1) + 1 \Rightarrow x = \frac{3^1 x - 2^0}{2^{3-1}} \Rightarrow x = -1 \notin N_+ \Rightarrow LS(n)$ does not exist, $\forall n \in N_+$

---

ii) $r = 2k + 1$, & $LS(n) = S(n) \Rightarrow x \neq 1, x \in N_+$

$$x = \frac{3^k x - 3^{k-1}(2^0) - 3^{k-2}(2^1) - 3^{k-3}(2^2) - \cdots - 3^1(2^{k-2}) - 3^0(2^{k-1})}{2^{r-k}} = 5 \in LS(n) \Rightarrow$$

$LS(n) = \{5,14,7,20,10\}, \forall n \in B$

---

iii. a) if $r = 2k$,

$r = 18 = 2(9) \Rightarrow$ Using eq1, $\forall n \in N_+/(A \cup B)$:

$$x = \frac{3^k x - 3^{k-1}(2^0) - 3^{k-2}(2^1) - 3^{k-3}(2^2) - \cdots - 3^1(2^{k-2}) - 3^0(2^{k-1})}{2^{r-k}} \Rightarrow$$

$(2^{r-k} - 3^k)x = -3^{k-1}(2^0) - 3^{k-2}(2^1) - 3^{k-3}(2^2) - \cdots - 3^1(2^{k-2}) - 3^0(2^{k-1})$

$(2^9 - 3^9)x = -3^8(2^0) - 3^7(2^1) - 3^6(2^2) - 3^5(2^3) - 3^4(2^4) - 3^3(2^5) - 3^2(2^6)$
$-3^1(2^7) - 3^0(2^8)$

$(-19171)x = -19171$

$x = 1 \Rightarrow LS(n) = \{1,2,1,2,1,2,1,2,1,2,1,2,1,2,1,2, 1,2\} = \{1,2\}. \forall n \in N_+/(A \cup B)$

iii. b) if $r = 2k, r = 18$, & $\forall n \in B \subset N_+$, then don't use eq1:

$\because LS(n) = \{17,50,25,74,37,110,55,164,82,41,122,61,182,91,272,136,68,34\} = Z, \forall n \in B \subset N_+$

$\therefore$ Murad Sequence has at least 3 different loops:

1) $LS(n) = \{1,2\}$, when $r = 2k$, $\forall n \in N_+/(A \cup B)$

2) $LS(n) = \{5, 14,7,20,10\}$, when $r = 5k$, $\forall n \in A \subset N_+$

3) $LS(n) = \{17,50,25,74,37,110,55,164,82,41,122,61,182,91,272,136,68,34\}$, when $r = 18k, \forall n \in B \subset N_+$

$$x = \frac{3^k x - 3^{k-1}(2^0) - 3^{k-2}(2^1) - 3^{k-3}(2^2) - \cdots - 3^1(2^{k-2}) - 3^0(2^{k-1})}{2^{r-k}}$$

$$x = \frac{3^k x - \sum_{i=0}^{k-1} 3^{k-1-i} 2^i}{2^{r-k}}$$

$$(2^{r-k} - 3^k)x = -\sum_{i=0}^{k-1} 3^{k-1-i} 2^i$$

$$x = \frac{-\sum_{i=0}^{k-1} 3^{k-1-i} 2^i}{2^{r-k} - 3^k}$$

$$x = \frac{S}{2^{r-k} - 3^k}$$

$$S = -\sum_{i=0}^{k-1} 3^{k-1-i} 2^i$$

Euler Perfect Box Best Way

Author: Taha M. Muhammad/USA Kurd Iraq

$$a, b, c, d, e, f, g \in N_+ \Leftrightarrow \text{Euler Perfect Box (EPB)}$$

Proof

A) Let $a < b < c$, & $a, b, c \in N_+$, $N_{Square} = N_S$ $\because g^2 = a^2 + f^2 \Rightarrow a^2 = g^2 - f^2 \dots (Pythagor)$ Add $g^2 + 4gf + 3f^2$ to both sides $\therefore a^2 + g^2 + 4gf + 3f^2 = g^2 - f^2 + g^2 + 4gf + 3f^2$ $a^2 + g^2 + 4gf + 3f^2 = 2g^2 + 4gf + 2f^2$ $a^2 + g^2 + 4gf + 3f^2 = (\sqrt{2}g + \sqrt{2}f)^2$ $t = \sqrt{2}g + \sqrt{2}f = \sqrt{2}(g + f), t \in N_+, \text{or } t \notin N_+$ $\therefore (g + f) = \frac{t}{\sqrt{2}}, t \neq \sqrt{2}$ $\therefore g = \frac{t}{\sqrt{2}} - f \Rightarrow f \in N_+, \& g \notin N_+ \Rightarrow \nexists EPT$ or $f = \frac{t}{\sqrt{2}} - g \Rightarrow f \notin N_+, \& g \in N_+ \Rightarrow \nexists EPT$	
B) if $a = b < c$ $\because d^2 = a^2 + b^2$ $\therefore d^2 = a^2 + a^2 \Rightarrow d^2 = 2a^2 \Rightarrow$ $d = \sqrt{2}\, a \Rightarrow d \notin N_+ \Rightarrow \nexists EPT$	
C) if $a = b = c$ $\because d^2 = a^2 + b^2$ $\therefore d^2 = a^2 + a^2 = 2a^2$ $\therefore d^2 = 2a^2 \Rightarrow d = a\sqrt{2} \Rightarrow d \notin N_+$ $\Rightarrow \nexists EPT$	
by $A, B, \& C \Rightarrow \nexists EPT$	

Taha's Geometry Theorem

"Given a circle and a central angle subtended by a chord, use a compass and straightedge to construct another circle with the same center, such that two connected chords subtend the same central angle and each chord is equal in length to the original chord."

Q: A circle center A, has chord BC. Draw by compass and straightedge another circle center A, such that EF = FG = BC

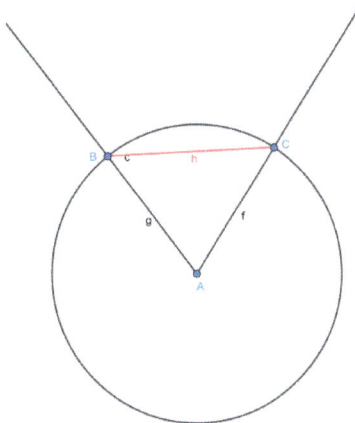

Given:

Circle c, angle BAC, and chord BC

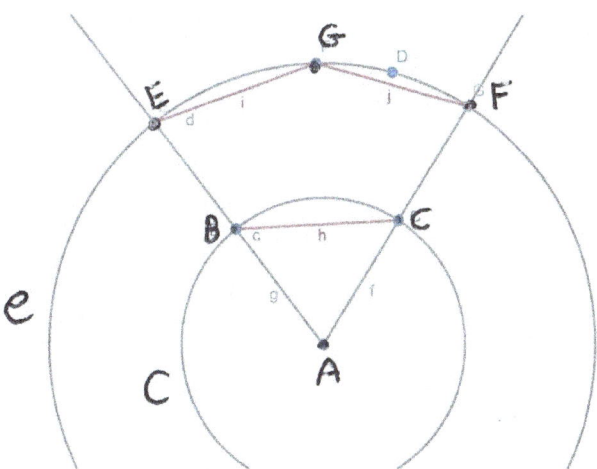

Required procedure:

Circle e such that EG = GF = BC

Procedure and proof:

Draw AN Bisector of $\angle BAC \Rightarrow (\angle 1 + \angle 2 = \angle 3 + \angle 4)$ & $\angle 8 = 90°$

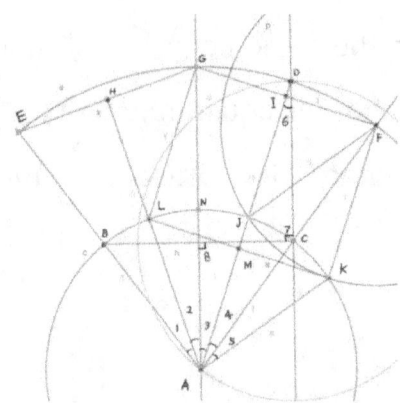

Draw line $CD \perp BC$ at point C $\Rightarrow (\angle 7 = 90°,$ since $\angle 8 = 90°) \Rightarrow$

$CD \parallel AG \Rightarrow \angle 6 = \angle 3$

Draw circle d (Center C, radius AC) $\cap$ line CD at point F

Draw AD $\cap$ circle c at point J

Draw circle e (Center A, radius AD) $\cap$ lines AC, AN, AB at F, G, E respectively

Draw EG, GF

$\because \angle BAC = \angle EAF \Rightarrow$ AG bisector of $\angle EAF \Rightarrow EG = GF \dots 1$

Draw AH Bisector of $\angle EAG \Rightarrow \angle 1 = \angle 2$

Draw circle p (Center F, radius FJ) $\cap$ circle c at K, draw FK, AK

$\triangle$ AJF, $\triangle$ ACD:

AC=AJ… (Radiuses of circle c)

AF=AD… (Radiuses of circle e)

$\angle 4 = \angle 4$ … (common angle)

$\therefore \triangle$ AJF $\equiv \triangle$ ACD… (Side, side, angle) $\Rightarrow FJ = DC = AC$… (Radiuses of circle d)

$\Rightarrow \angle 4 = \angle 6 \Rightarrow (\angle 4 = \angle 3) \Rightarrow$ AD Bisector of $\angle GAF \Rightarrow \left(GI = IF = \frac{1}{2}GF\right)$ & $\angle FIM = $

$90°\dots 3$

$\therefore$ FK = FJ = DC = AC = AJ $\Rightarrow$ AJFK rhombus $\Rightarrow \angle 4 = \angle 5 \Rightarrow$ FK $\parallel$ AJ $\Rightarrow$ KF $\parallel$ IM ... 5

$\because \angle 1 + \angle 2 = \angle 3 + \angle 4 = 2\angle 1 = 2\angle 2 = 2\angle 3 = 2\angle 4 = \angle 1 = \angle 2 = \angle 3 = \angle 4 = \angle 5$

$\therefore \angle 1 + \angle 2 + \angle 3 + \angle 4 = \angle 2 + \angle 3 + \angle 4 + \angle 5 = $ BC = LK ... 2

$\because \angle 2 + \angle 3 = \angle 4 + \angle 5 \Rightarrow$ AD Bisector of$\angle$LAK $\Rightarrow \left(LM = MK = \dfrac{1}{2} LK \right)$ & $\angle$IMK

$\qquad = 90° ... 4$

$\therefore$ [MK $\parallel$ IF... (By 3, 4), & $KF \parallel$ IM ... (By 5)] $\Rightarrow$ IFKM is a rectangle $\Rightarrow$ IF = MK = $\dfrac{1}{2}$ LK = $\dfrac{1}{2}$ GF

$\therefore$ LK = GF ... 6

$\therefore$ BC = GF ... (By 2,6)

$\because$ EG = GF ... 1

$\therefore$ EG = GF = BC ... (Substitution)

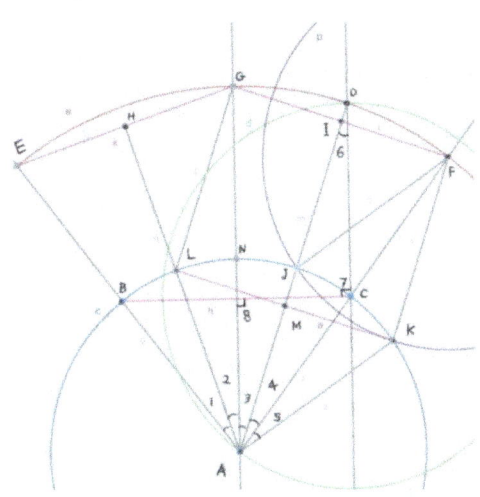

09- Pythagorean Theorem

	Pythagorean Theorem Author: Taha Muhammad USA Kurd	

$$is\ a^2 + b^2 = c^2?$$
Proof

ABC Right Triangle at C *c is hypotenuse* $a < b,$ or $a = b'$ $c < a + b, a < c + b, b < a + c$ (Triangle Property (T.P.)) $a, b, c, r, g, h, u, v, j, k, m, n \in R_+$	

let $c^2 \neq a^2 + b^2 \Rightarrow$

A) if $c^2 > a^2 + b^2 \Rightarrow c^2 = a^2 + b^2 + r$

i) if $r = 2ab \Rightarrow c^2 = a^2 + b^2 + 2ab = (a + b)^2 \Rightarrow c = a + b \dots$ Contradiction to

$c < a + b \dots$ (Tri. Pro.)

ii) if $r \neq 2ab \Rightarrow c^2 = a^2 + b^2 + r \Rightarrow$

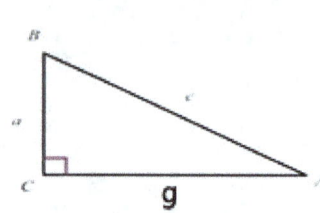 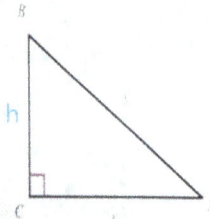 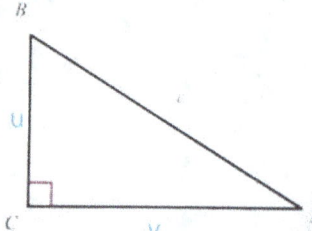

$[c^2 = a^2 + 2ag + g^2 \Rightarrow c^2 = (a + g)^2 \Rightarrow c = a + g \dots$ Contradiction to $c < a + g,$

or $c^2 = h^2 + 2hb + b^2 \Rightarrow c^2 = (h + b)^2 \Rightarrow c = h + b \dots$ Contradiction to $c < h + b,$

or $c^2 = u^2 + 2uv + v^2] \Rightarrow c^2 = (u + v)^2] \Rightarrow c = u + v \dots$ Contradiction to $c < u + v] \Rightarrow$

$r = 2ab \Rightarrow c = a + b \dots$ Contradiction to $c < a + b \dots$ (Tri. Pro.)

$\therefore$ by i & *ii*: $c^2 \ngtr a^2 + b^2 \dots$ (A)

B) if $c^2 < a^2 + b^2 \Rightarrow c^2 = a^2 + b^2 - r$

i) if $r = 2ab \Rightarrow c^2 = a^2 - 2ab + b^2 = (a - b)^2 \Rightarrow$

$c = a - b \Rightarrow c = 0$ or $c \in R_{(-)} \dots$ Contradiction to $c \in R_+$

ii) if $r \neq 2ab \Rightarrow c^2 = a^2 + b^2 - r \Rightarrow$

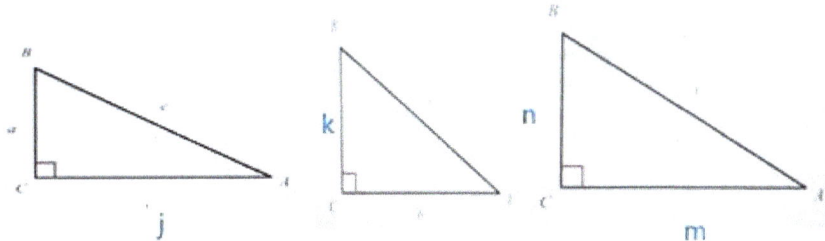

$c^2 = a^2 - 2aj + j^2 = (a - j)^2 \Rightarrow c = a - j \Rightarrow c < a \dots Contradiction\ to\ c > a$

or $c^2 = k^2 - 2kb + b^2 = (k - b)^2 \Rightarrow c = k - b \Rightarrow c < k \Rightarrow \cdots Contradiction\ to\ c > k$

or $c^2 = m^2 - 2mn + n^2 = (m - n)^2 \Rightarrow c = m - n \Rightarrow c < m \dots Contradiction\ to\ c > m] \Rightarrow$

$r = 2ab \Rightarrow c = a - b \Rightarrow c = 0$ or $c \in R_{(-)} \dots$ Contradiction to $c \in R_+$

$\therefore$ by i & ii: $c^2 \not< a^2 + b^2 \dots$ (B)

$\therefore c^2 = a^2 + b^2 \dots$ by A & B

10- *Fermat's Last Theorem*

Fermat's Last Theorem 1ˢᵗ Way
Abstract

$[is\ a^n + b^n = c^n?, a < b < c, \& \ a, b, c, n \ \epsilon N_+, n > 2] \Leftrightarrow [Fermat's\ Last\ Theorem]$

■ *Taha's Coefficient Fact1 (TCF1):* $a + b = c \Rightarrow aa^{r-1} + bb^{r-1} \neq cc^{r-1} \Rightarrow$ $a^r + b^r \neq c^r, when\ a^r \neq b^r \neq c^r, r \neq 1, r\ \epsilon N_+.$
Example: $4+5=9 \Rightarrow 4(4^2) + 5(5^2) \neq 9(9^2) \Rightarrow 189 \neq 729$

■ *Taha's Coefficient Fact2 (TCF2):* $a + b \neq c \Rightarrow aa^{r-1} + bb^{r-1} \neq cc^{r-1} \Rightarrow$ $a^r + b^r \neq c^r, when\ a^r \neq b^r \neq c^r, r \neq 1, r\ \epsilon N_+.$
Example: $4+5\neq6 \Rightarrow 4(4^2) + 5(5^2) \neq 6(6^2) \Rightarrow 189 \neq 216$

■ *Taha's* (N_+) *& Three Sided Geometric Shapes Fact (TNGSF)*
$N_+ = (Right\ Triangl\ S_1) \cup (Acute\ Triangle\ S_2) \cup (Obtuse\ Triangle\ S_3)$
$\cup (Segment\ S_4 : c = a + b) \cup (Segment\ S_5 : c > a + b)$

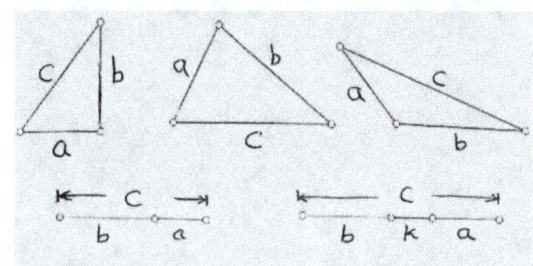

1) *if* $a = b$ *in all shapes*	2) *if* $a = b = c$ *in Acute Triangle Only*
Proof:	*Proof:*
is $a^n + b^n = c^n?$	*is* $a^n + b^n = c^n?$
let $a^n + b^n = c^n$	*let* $a^n + b^n = c^n$
$a^n + a^n = c^n$	$\therefore a^n + a^n = a^n$
$2a^n = c^n$	$a^n = 0 \Rightarrow a = 0 \dots Contradiction\ to\ a \in N_+$
$\therefore c = \sqrt[n]{2}\ \ a \notin N_+ \dots Contradiction\ to\ c \in N+$	$\therefore a^n + b^n \neq c^n$
$\therefore a^n + b^n \neq c^n$	

3) *if $a < b < c$, for all three-sided Geometric shapes*

S_1: *Right Triangle*

is $a^n + b^n = c^n$? when $a < b < c$? $\because a^2 + b^2 = c^2 \Rightarrow a^{n-2} a^2 + b^{n-2} b^2 \neq c^{n-2} c^2 \dots (TCF)$ $\Rightarrow a^n + b^n \neq c^n$	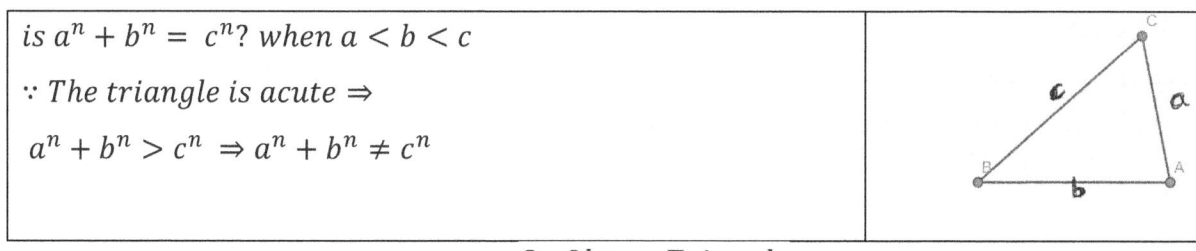

S_2: *Acute Triangle*

is $a^n + b^n = c^n$? when $a < b < c$ $\because$ The triangle is acute $\Rightarrow$ $a^n + b^n > c^n \Rightarrow a^n + b^n \neq c^n$	

S_3: *Obtuse Triangle*

is $a^n + b^n = c^n$? when $a < b < c$ $\because$ The triangle is Obtuse $\Rightarrow$ $a^n + b^n < c^n \Rightarrow a^n + b^n \neq c^n$	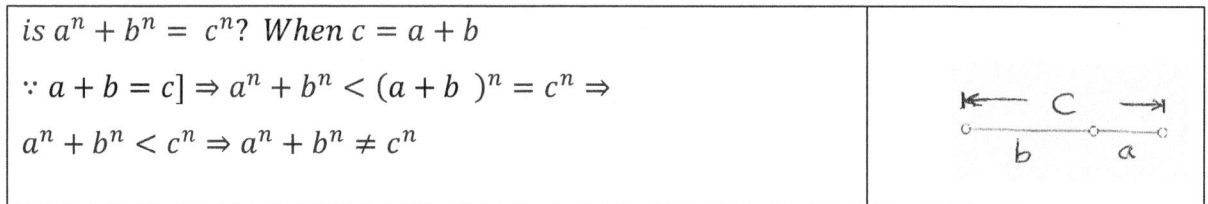

S_4: $c = a + b$

is $a^n + b^n = c^n$? When $c = a + b$ $\because a + b = c] \Rightarrow a^n + b^n < (a + b)^n = c^n \Rightarrow$ $a^n + b^n < c^n \Rightarrow a^n + b^n \neq c^n$	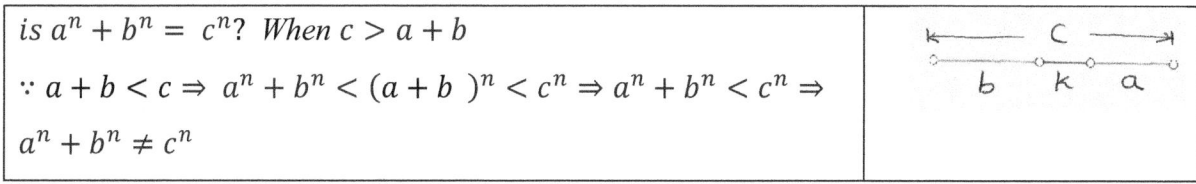

S_5: $c > a + b$

is $a^n + b^n = c^n$? When $c > a + b$ $\because a + b < c \Rightarrow a^n + b^n < (a + b)^n < c^n \Rightarrow a^n + b^n < c^n \Rightarrow$ $a^n + b^n \neq c^n$	

Conclusion: $a^n + b^n \neq c^n \dots \forall n \in S_1 U S_2 U S_3 U S_4 U S_5 = N_+, when\ n \geq 3$

Both Fermat's by Compass & Straightedge 2nd Way

Title: A New Solution to Fermat's Last Theorem, and General Case – From an Independent Mathematician

"Taha M. Muhammad"

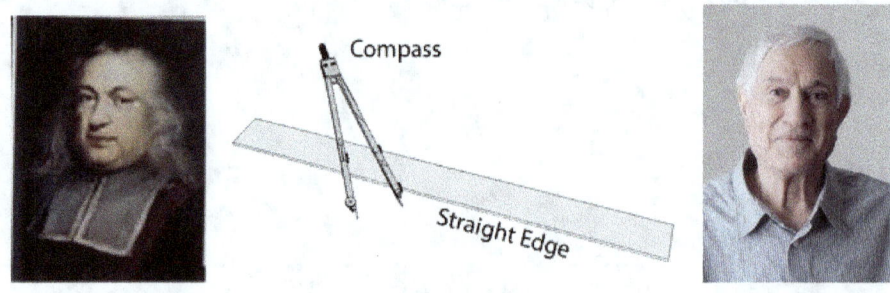

Fermat's Last Theorem 2ⁿᵈ Way
Shape 1: Right Triangle ABC

$$is\ a^n + b^n\ =\ c^n?$$

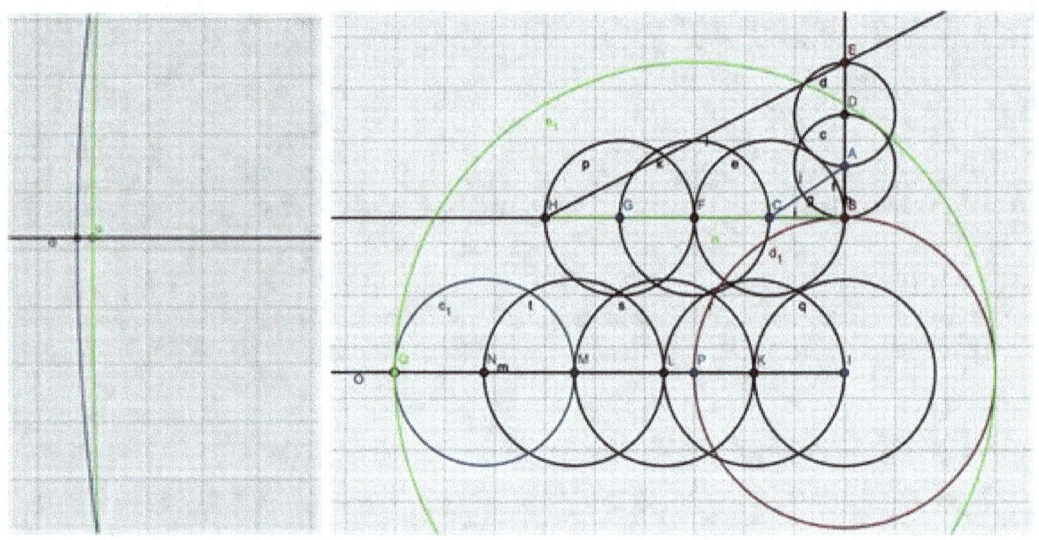

Proof :

$a^n + b^n\ =\ a^{n-1}\,a^1 + b^{n-1}\,b^1$

$c^n\ =\ c^{n-1}c^1$

$let\ a^{n-1} = 3, b^{n-1} = 4,\ c^{n-1} = 5$

$\therefore 3a =\ a^n, 4b =\ b^n, 5c = c^n$

$\because BE + BH = IP + PQ = IQ > IO$

$\therefore\ a^n + b^n > c^n$

$\therefore a^n + b^n \neq c^n$

$$is\ a^n + b^n = c^n?$$

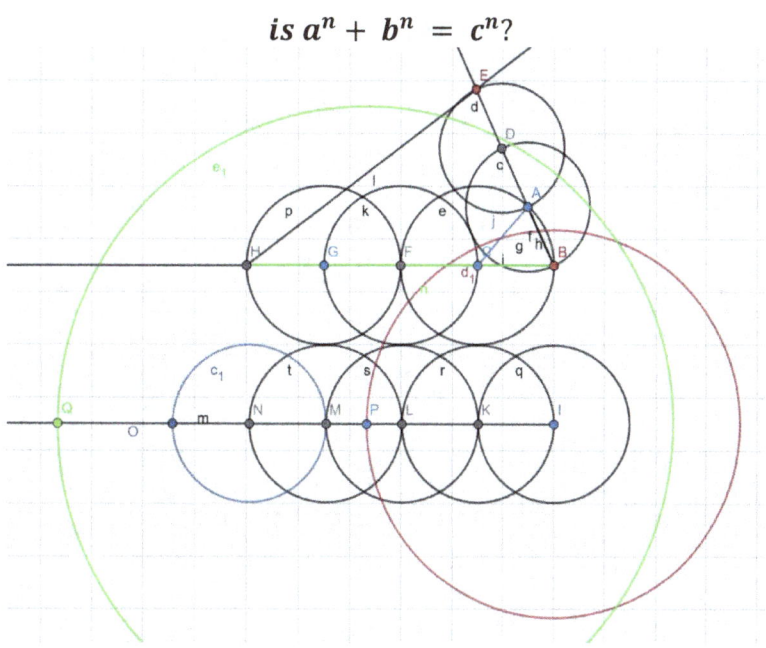

Proof :

$$a^n + b^n = a^{n-1}\,a^1 + b^{n-1}\,b^1$$

$$c^n = c^{n-1}c^1$$

$$let\ a^{n-1} = 3, b^{n-1} = 4,\ c^{n-1} = 5$$

$$\therefore 3a = a^n, 4b = b^n, 5c = c^n$$

$$\because BE + BH = IP + PQ = IQ > IO$$

$$\therefore\ a^n + b^n > c^n$$

$$\therefore a^n + b^n \neq c^n$$

$$\textbf{is } a^n + b^n = c^n?$$

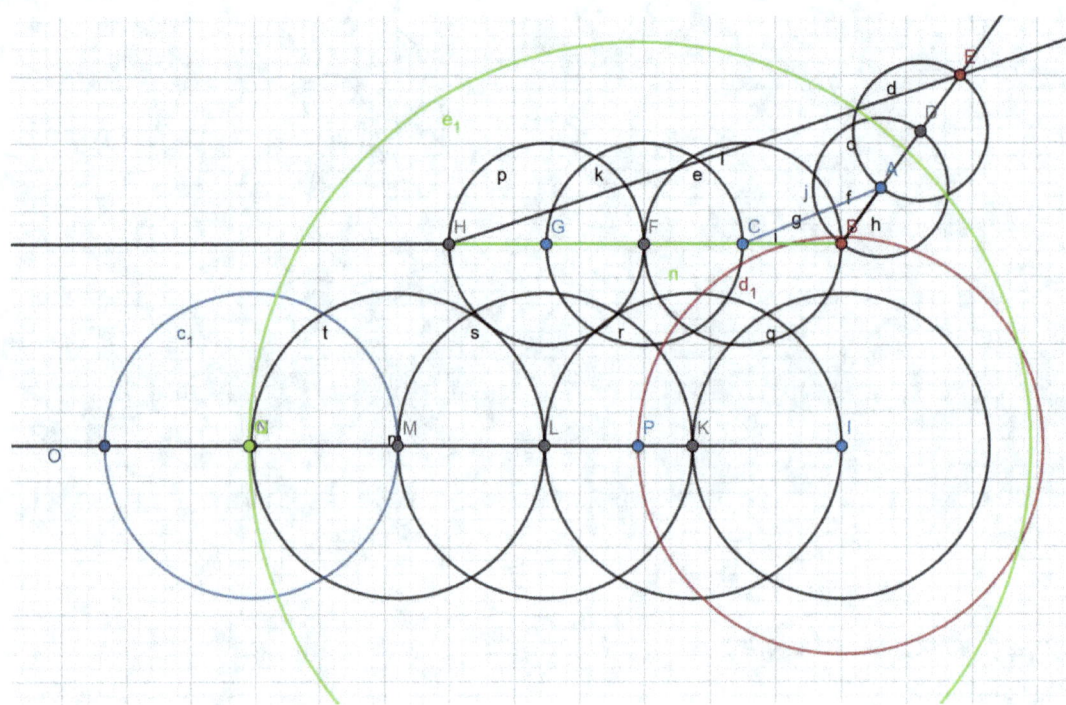

Proof :

$a^n + b^n = a^{n-1} a^1 + b^{n-1} b^1$

$c^n = c^{n-1} c^1$

let $a^{n-1} = 3, b^{n-1} = 4, c^{n-1} = 5$

$\therefore 3a = a^n, 4b = b^n, 5c = c^n$

$\because BE + BH = IP + PQ = IQ < IO$

$\therefore a^n + b^n < c^n$

$\therefore a^n + b^n \neq c^n$

$$is\ a^n + b^n = c^n?$$

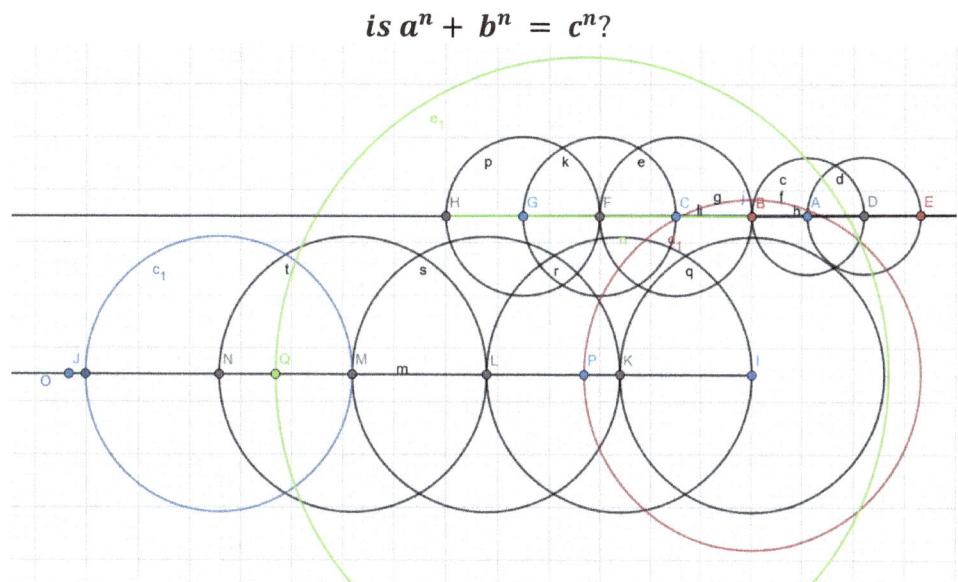

Proof :

$$a^n + b^n = a^{n-1}\,a^1 + b^{n-1}\,b^1$$

$$c^n = c^{n-1}c^1$$

$$let\ a^{n-1} = 3, b^{n-1} = 4,\ c^{n-1} = 5$$

$$\therefore 3a = a^n, 4b = b^n, 5c = c^n$$

$$\because BE + BH = IP + PQ = IQ < IO$$

$$\therefore a^n + b^n < c^n$$

$$\therefore a^n + b^n \neq c^n$$

Shape 5: *Segment* $c = a + k + b$

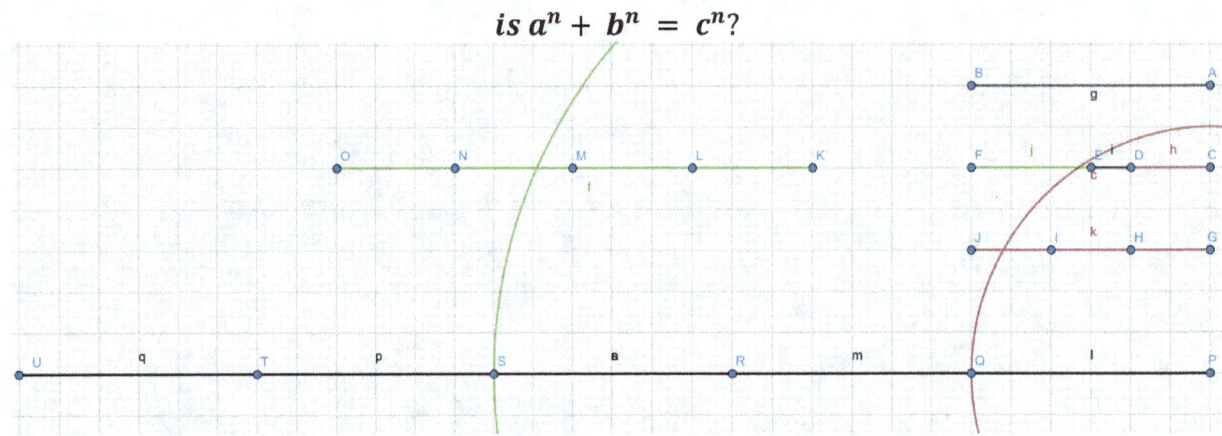

is $\boldsymbol{a^n + b^n = c^n}$?

Proof :

$a^n + b^n = a^{n-1} a^1 + b^{n-1} b^1$

$c^n = c^{n-1} c^1$

$let\ a^{n-1} = 3, b^{n-1} = 4,\ c^{n-1} = 5$

$\therefore 3a = a^n, 4b = b^n, 5c = c^n$

$AB = c, DC = a, ED = k, EF = b$

$GJ = a^n, KO = b^n$

$\therefore GJ + KO = PQ + QS = a^n + b^n$

$\&\ PU = c^n$

$\because RV < RU$

$\therefore a^n + b^n < c^n \Rightarrow a^n + b^n \neq c^n$

Note

1) Shape 1 U Shape 2 U Shape 3 U Shape 4 U Shape 5 = N_+
$\therefore a^n + b^n \neq c^n\ for\ n \in\ N_+$

2) Same way Fermat's General Case, but $a^n + b^m \neq c^z$ & in some cases $a^n + b^m = c^z$

Fermat's Last Theorem 3rd Way

Abstract

$[is\ a^n + b^n = c^n?, a < b < c, \& \ a,b,c,n \in N_+, n > 2] \Leftrightarrow [Fermat's\ Last\ Theorem]$

■ *Taha's Coefficient Fact1 (TCF1):* $a + b = c \Rightarrow aa^{r-1} + bb^{r-1} \neq cc^{r-1} \Rightarrow$
$a^r + b^r \neq c^r, when\ a^r \neq b^r \neq c^r, r \neq 1, r \in N_+.$
Example: $4+5=9 \Rightarrow 4(4^2) + 5(5^2) \neq 9(9^2) \Rightarrow 189 \neq 729$

■ *Taha's Coefficient Fact2 (TCF2):* $a + b \neq c \Rightarrow aa^{r-1} + bb^{r-1} \neq cc^{r-1} \Rightarrow$
$a^r + b^r \neq c^r, when\ a^r \neq b^r \neq c^r, r \neq 1, r \in N_+.$
Example: $4+5 \neq 6 \Rightarrow 4(4^2) + 5(5^2) \neq 6(6^2) \Rightarrow 189 \neq 216$

■ *Taha's (N_+) & Three Sided Geometric Shapes Fact (TNGSF)*
$N_+ = (Right\ Triangl\ S_1) \cup (Acute\ Triangle\ S_2) \cup (Obtuse\ Triangle\ S_3)$
$\cup (Segment\ S_4: c = a + b) \cup (Segment\ S_5: c > a + b)$

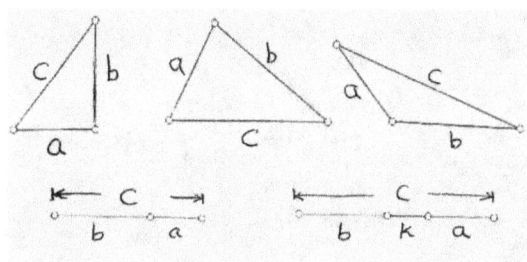

1) *if $a = b$ in all shapes*	2) *if $a = b = c$ in Acute Triangle Only*
Proof:	*Proof:*
is $a^n + b^n = c^n$?	*is $a^n + b^n = c^n$?*
let $a^n + b^n = c^n$	*let $a^n + b^n = c^n$*
$a^n + a^n = c^n$	$\therefore a^n + a^n = a^n$
$2a^n = c^n$	$a^n = 0 \Rightarrow a = 0\ ... Contradiction\ to\ a \in N_+$
$\therefore c = \sqrt[n]{2}$	$\therefore a^n + b^n \neq c^n$
$a \notin N_+ ... Contradiction\ to\ c \in N_+$	
$\therefore a^n + b^n \neq c^n$	

3) *if $a < b < c$, for all 3-sided Geometric shapes*

is $a^n + b^n = c^n$?

1st side: $(a^n + b^n)\ times\ b \Rightarrow b(a^n + b^n), \&$

2nd side: (c^n) *times* $c \Rightarrow c(c^n)$

1) *either* $b(a^n + b^n) = c(c^n), \& \, b < c \Rightarrow (a^n + b^n) > (c^n) \Rightarrow a^n + b^n \neq c^n$

2) *or* $b(a^n + b^n) \neq c(c^n) \Rightarrow$

i) $b(a^n + b^n) < c(c^n)$

$ba^n + b^{n+1} < c^{n+1} \dots (*)$

$\because a < b,$ *then change* (ba^n) *to* (aa^n) *in LHS of* $(*)$

$\therefore aa^n + b^{n+1} < c^{n+1} \Rightarrow a^{n+1} + b^{n+1} \neq c^{n+1} \Rightarrow a^n + b^n \neq c^n \dots TCF2$

ii)*or* $b(a^n + b^n) > c(c^n) \Rightarrow ba^n + b^{n+1} > c^{n+1} \dots (**),$ *then change* (ba^n) *to* (aa^n)

$\therefore LHS \ of \ (**) = \ ba^n + b^{n+1} \Rightarrow aa^n + b^{n+1} = a^{n+1} + b^{n+1}$

$\& \ RHS \ of \ (**) = \ c^{n+1}.$

There are 3 posibilities between LHS & RHS:

A) $a^{n+1} + b^{n+1} > c^{n+1} \Rightarrow a^{n+1} + b^{n+1} \neq c^{n+1} \Rightarrow a^n + b^n \neq c^n \dots TCF2$

B) $a^{n+1} + b^{n+1} < c^{n+1} \Rightarrow a^{n+1} + b^{n+1} \neq c^{n+1} \Rightarrow a^n + b^n \neq c^n, \dots TCF2$

C) *or* $a^{n+1} + b^{n+1} = c^{n+1} \Rightarrow a^n + b^n \neq c^n, \dots TCF1$

Conclusion: $a^n + b^n \neq c^n,$ *for all* $n \, \epsilon \, N_+, n > 2.$

3) *if $a < b < c$, for all 3-sided Geometric shapes*

S_1: *Right Triangle*

is $a^n + b^n = c^n$? $\because S_1$ *is a Right Triangle* $\Rightarrow a^2 + b^2 = c^2 \Rightarrow$ $a^{n-2} a^2 + b^{n-2} b^2 \neq c^{n-2} c^2$ ...(TCF1) $\Rightarrow a^n + b^n \neq c^n$	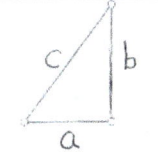

S_2: *Acute Triangle*

is $a^n + b^n = c^n$? $\because S_2$ *is Acute Triangle* $\Rightarrow$ *Right Triangle sides*: $a, b, d \Rightarrow$ $a^2 + b^2 = d^2 > c^2 \Rightarrow a^2 + b^2 > c^2$ $\Rightarrow a^n + b^n \neq c^n$ ...(TCF1)	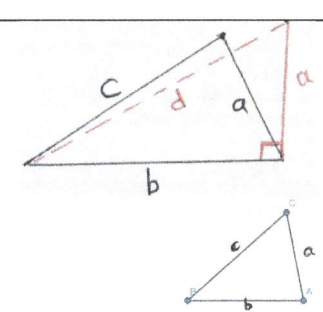

S_3: *Obtuse Triangle*

is $a^n + b^n = c^n$? $\because S_3$ *is Obtuse Triangle* $\Rightarrow$ *Right Triangle sides*: a, b, d $a^2 + b^2 = d^2 < c^2 \Rightarrow a^2 + b^2 < c^2 \Rightarrow$ $a^n + b^n \neq c^n$ ...(TCF1)	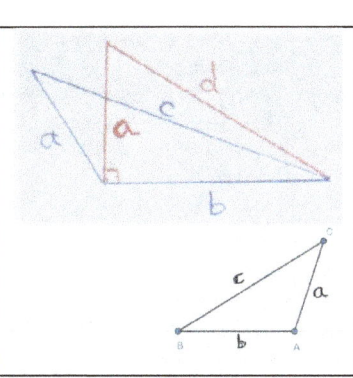

S_4: $c = a + b$

is $a^n + b^n = c^n$? When $c = a + b$ *in S_4*: $a + b = c \Rightarrow a^n + b^n < (a + b)^n = c^n \Rightarrow$ $a^n + b^n < c^n \Rightarrow a^n + b^n \neq c^n$	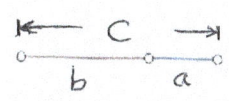

S_5: $c > a + b$

is $a^n + b^n = c^n$? When $c > a + b$ *in S_5*: $a + b < c \Rightarrow a^n + b^n < (a + b)^n < c^n \Rightarrow$ $a^n + b^n < c^n \Rightarrow a^n + b^n \neq c^n$	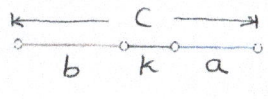

Conclusion: $a^n + b^n \neq c^n \ \forall \ n \in (S_1 U S_2 U S_3 U S_4 U S_5) = N_+,$ *when* $n > 2$

Fermat's Last Theorem 5th Way

Abstract

$$is \ a^n + b^n = c^n?, a, b, c, n \epsilon N_+, n > 2 \Leftrightarrow Fermat's \ Last \ Theorem$$

Proof:

$$let \ a^n + b^n = c^n$$

1) $if \ a = b$	2) $if \ a = b = c$
$\therefore a^n + a^n = c^n$	$\therefore a^n + a^n = a^n$
$2a^n = c^n$	$a^n = 0 \Rightarrow a = 0 \ldots Contradiction \ to \ a \in N_+$
$c = (\sqrt[n]{2} \ a) \notin N_+ \ldots Contradiction \ to \ c \in N_+$	$\therefore a^n + b^n \neq c^n$
$\therefore a^n + b^n \neq c^n$	

$if \ a < b < c, then$:

$Case \ 1) \ a + b = c$	$Case \ 2) \ a + b < c$	$or \ Case \ 3) \ a + b > c$

$Case 1) \ a + b = c \Rightarrow (a + b)^n = c^n \Rightarrow a^n + b^n < (a + b)^n = c^n \Rightarrow a^n + b^n \neq c^n.$

$Case 2) \ a + b < c \Rightarrow (a + b)^n < c^n \Rightarrow a^n + b^n < (a + b)^n < c^n \Rightarrow a^n + b^n \neq c^n.$

$Case 3) \ a + b > c \Rightarrow (a + b)^n > c^n \Rightarrow a^n + b^n < (a + b)^n > c^n \Rightarrow$

$i) \ a^n + b^n = c^n, ii) \ a^n + b^n > c^n, iii) \ a^n + b^n < c^n$

$i) \because \ a + b = c \Rightarrow (a + b)^n = c^n \Rightarrow a^n + k + b^n = c^n, k \epsilon N_+ \Rightarrow$

$\quad a^n + b^n = c^n \Leftrightarrow k = 0 \ when \ a = 0, or \ b = 0, \& \ a + b = c.$

$\therefore a^n + b^n \neq c^n \Leftrightarrow k \neq 0 \ when \ a \neq 0, and \ b \neq 0, \& \ a + b \neq c.$

$ii) \ a^n + b^n > c^n \Rightarrow a^n + b^n \neq c^n.$

$iii) \ a^n + b^n < c^n \Rightarrow a^n + b^n \neq c^n.$

$Conclusion$: $a^n + b^n \neq c^n \ for \ all \ n \ \epsilon N_+$

11- Fermat's General Case

Fermat's General Case 1st Way

Abstract

$$[a^n + b^m = c^z? \ \& \ a, b, c, n, m, z \in N_+] \Leftrightarrow [Fermat's \ General \ Case]$$

Taha's (N_+) & Three Sided Geometric Shapes Fact

$N_+ = (Right \ Triangls) \cup (Acute \ Triangles) \cup (Obtuse \ Triangles)$
$\cup (Seqments \ c = a + b) \cup (Seqments \ c > a + b)$

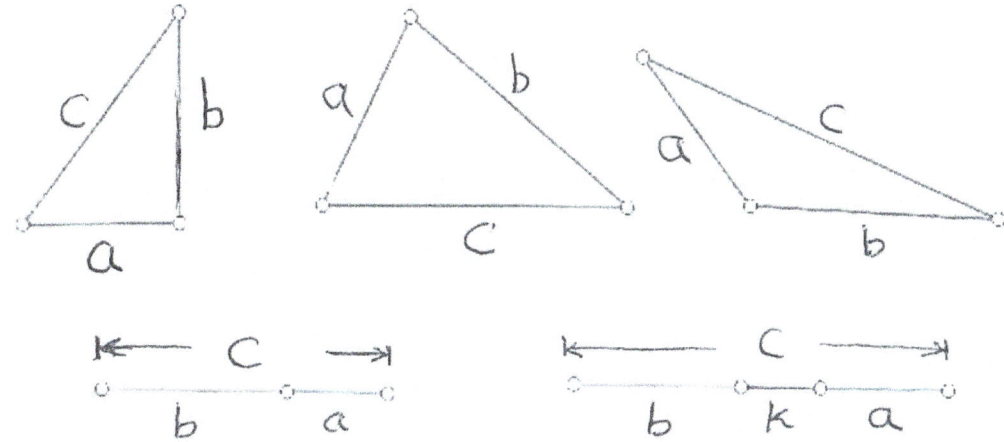

Control of Exponents n, m, and z

Relationship among exponents n, m, and z	
i	*Ii*
$n = m < z$	$n > m > z$
$n = z < m$	$n > z > m$
$m = z < n$	$m > n > z$
$m = n > z$	$m > z > n$
$z = n > m$	$z > n > m$
$z = m > n$	$z > m > n$

Shape1: *Right Triangle*

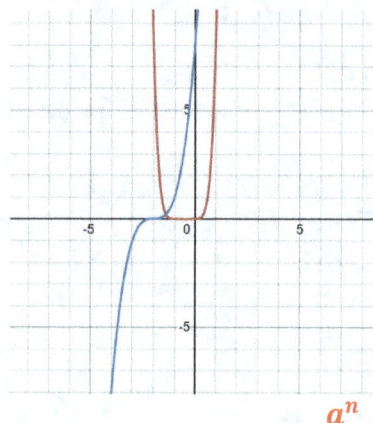 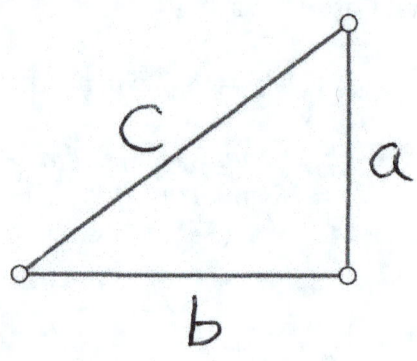

$$a^n + b^m \neq c^z$$

A) is $a^n + b^m = c^z$? when $n \leq m < z$ & $a \leq b < c$

let $a^n + b^m = c^z$

$a^n + b^m = c^{m+r}$, & $r \in N_+$

$a^n + b^m > b^{m+r}$

$a^n > b^{m+r} - b^m$

$a^n > b^m(b^r - 1)$, & $(b^r - 1) > 1$

$a^n > b^m$ ... contradiction to $a^n < b^m$

$\therefore a^n + b^m \neq c^z$

B) is $a^n = c^m + b^z$? when $n = m < z$, & $a < b < c$

let $a^n = c^m + b^z$ ... eq1

$\because a^n + b^n \neq c^n$ ... (Fermat's Last Theorem in Right Triangle)

$\therefore$ i) $c^n > a^n + b^n$, or ii) $c^n < a^n + b^n$:

i) if $c^n > a^n + b^n$

$c^n = a^n + b^n + k$ ... eq2, $k \in N_+$,

$\because a^n = c^m + b^z$ ... eq1

$c^n + a^n = a^n + b^n + k + c^m + b^z$ ... (by adding eq1 to eq2) $\Rightarrow$

$c^n - c^m = b^n + k + b^z$

$c^n - c^n = b^n + k + b^z$ ... (because $n = m$)

$\therefore 0 = b^n + k + b^z$ contradiction to $b^n + k + b^z > 0$

$\therefore a^n \neq c^m + b^z$

$\therefore$ the same way $b^n \neq c^m + a^z$.

ii) if $c^n < a^n + b^n$ ... ineq1

$\therefore c^n < a^n + b^n < a^m + b^z$

$\therefore c^n < a^m + b^z$

$\therefore a^n < c^m + b^z$ because $a^n < c^n$ & $a^m < c^m$

$\therefore a^n \neq c^m + b^z$

$\therefore$ the same way $b^n \neq c^m + a^z$

Shape2: Acute Triangle

$$n \leq m < z \ \& \ a \leq b < c$$

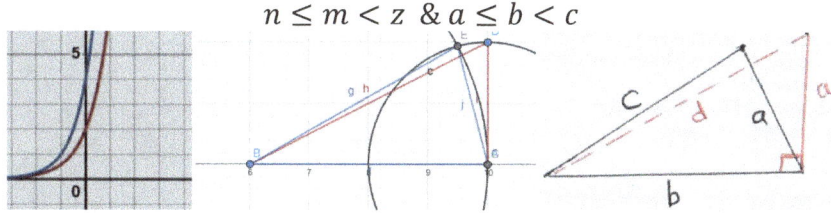

A) is $a^n + b^m = c^z$?, when $n < m < z \ \& \ a < b < c$?

let $a^n + b^m = c^z$

$\Rightarrow a^n + b^m > b^z \dots$ because $b^z < c^z$

$\Rightarrow a^n + b^m > b^{m+r}, r \in N_+$

$\Rightarrow a^n + b^m > b^m b^r$

$\Rightarrow a^n > b^m b^r - b^m$

$\Rightarrow a^n > b^m(b^r - 1), \ \& \ (b^r - 1) > 1 \dots$ Contardiction to $a^n < b^m(b^r - 1)$

$\therefore a^n + b^m \neq c^z$

B) is $a^n + b^m = c^z$? when $a = b < c, \ \& \ n = m < z$

let $a^n + b^m = c^z \Rightarrow$ $a^n + a^n > a^z$ $a^n + a^n > a^{n+r}, \ \& \ r \in N_+$ $a^n > a^{n+r} - a^n$ $\therefore a^n > a^n(a^r - 1), \ \& \ (a^r - 1 > 1)$ $a^n > a^n \Rightarrow a > a \dots$ Contradiction to $a = a$ $\therefore a^n + b^m \neq c^z$	$a^n + a^n < c^z$

C) is $a^n + b^m = c^z$? when $(a = b = c)$, & $n = m < z$, ... *Shape 2*

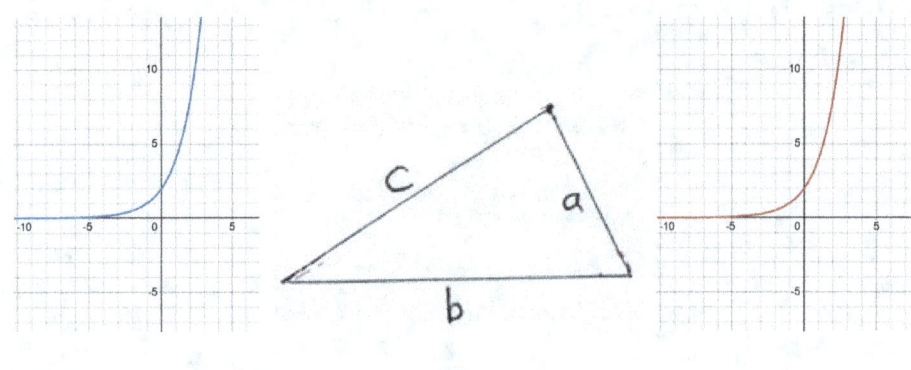

$$\mathbf{2a^n = 2(a^{n+1})}$$

let $a^n + b^m = c^z \dots eq1 \Rightarrow (a^n + a^n = a^z) \Rightarrow 2a^n = a^z \dots eq2)$

i) if $(a = 2, \& 1 = z - n) \Rightarrow$ LHS of eq2 = RHS of eq2 = 2

$\therefore a^n + b^m = c^z \Leftrightarrow 2(2^n) = 2^z$

$n = x$	$z = y$	a	b	c	
1	2	2	2	2	
2	3	2	2	2	
3	4	2	2	2	
4	5	2	2	2	
5	6	2	2	2	
6	7	2	2	2	
n	$n+1$	2	2	2	

Exapmles: $(2^1 + 2^1 = 2^2), (2^2 + 2^2 = 2^3), (2^3 + 2^3 = 2^4), \dots, (2^{17} + 2^{17} = 2^{18}), \dots$

ii) if $a \neq 2 \Rightarrow 2 = a^{z-n} \Rightarrow a \notin N_+ \dots$ Contradiction to $(a \in N_+)$

$\therefore a^n + b^m \neq c^z \ \forall a, b, c \in [N_+ - \{2\}] = \{1,3,4,5,6,7,8, \dots\}$

Exapmles: When $z > n = m$; $3^n + 3^n \neq 3^z$, $4^n + 4^n \neq 4^z$, $\dots$

Shape3: *Obtuse Triangle*

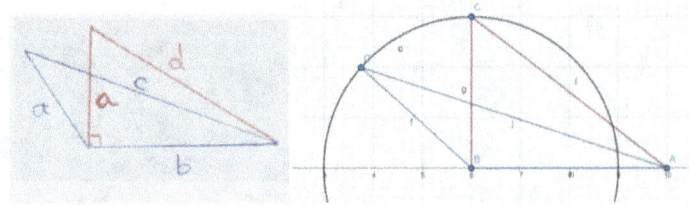

A) is $a^n + b^m = c^z$, $n \leq m < z$ & $a < b < c$? i) let $a^n + b^m = 3^3 + 4^3 = 91$, $c^z = 6^4 = 1296$] $\Rightarrow$ $a^n + b^m < c^z$ ...when $n = m < z$ ii) let $a^n + b^m = 3^3 + 4^4 = 283$, $c^z = 4^5 = 1024$] $\Rightarrow$ $a^n + b^m < c^z$ ...when $n < m < z$	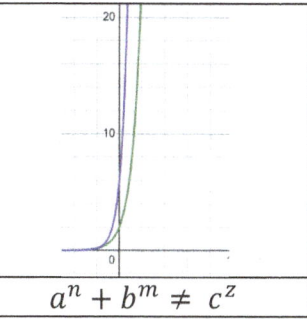 $a^n + b^m \neq c^z$

C) is $a^n + b^m = c^z$? when $a = b$ & $n = m < z$

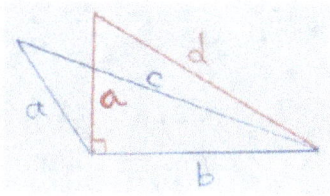

let $a^n + b^m = c^z$

$\therefore a^n + a^n = c^z$ because $(a = b)$ & $(n = m) \Rightarrow 2a^n = c^z \Leftrightarrow a = c = 2, \& z = n + 1$

$\Rightarrow a = b = c = 2 \Rightarrow$ Acute Δ ... contradiction to Obtuse Δ

$\therefore a^n + b^m \neq c^z$

Shape 4: $c = a + b$

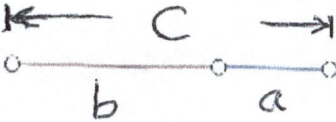

A) is $a^n + b^m = c^z$? when $a < b < c$, & $n = m < z$

let $a^n + b^m = c^z$

$\therefore a^n + b^m = (a + b)^z > a^z + b^z$ ... (by substitution)

$\therefore a^n + b^n > a^z + a^z$ ... contradiction to $a^n + b^n < a^z + a^z$ when $n < z$

$\therefore a^n + a^m \neq c^z$

B) is $a^n + b^m = c^z$? when $a = b < c$, & $n < m < z$

let $a^n + b^m = c^z$

$\therefore a^n + a^m = (a + b)^z > a^z + a^z$ ... (by substitution)

$\therefore a^n + a^m > a^z + a^z$ ... contradiction to $a^n + a^m < a^z + a^z$, when $n < m < z$

$\therefore a^n + b^m \neq c^z$.

C) is $a^n + b^m = c^z$? when $a < b < c$, & $n < m < z$

let $a^n + b^m = c^z$

$\therefore a^n + b^m = (a + b)^z > a^z + b^z \dots (by\ substitution)$

$\therefore a^n + b^m > a^z + b^z \dots contradiction\ to\ a^n + b^m < a^z + b^z, when\ n < m < z$

$\therefore a^n + b^m \neq c^z.$

D) $is\ a^n + b^m = c^z?\ when\ a = b < c,\ \&\ n = m < z$

$let\ a^n + b^m = c^z \Rightarrow a^n + a^n = (a + b)^z = (a + a)^z > (2\,a)^z \Rightarrow$

$a^n + a^n = 2a^n > (2\,a)^z \dots contradiction\ to\ 2a^n < (2\,a)^z \Rightarrow$

$a^n + b^m \neq c^z$

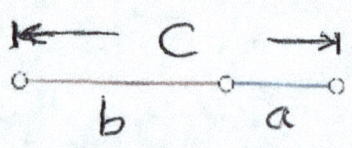

$but\ a^n + b^m = c^z?\ when\ a = b < c,\ \&\ n = m < z$

$if\ 2\,(4)^n = 8^z \dots a = b = 4, c = 8, \&\ look\ at\ x\ \&\ z\ column\ below$

$$2(4^x) = 8^y$$

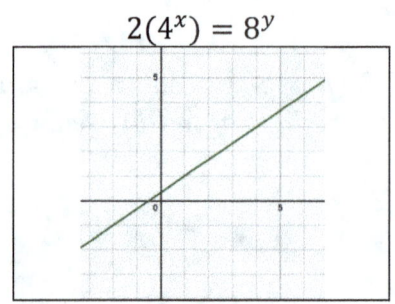

$n = x \downarrow$	$z = y$	$n - z$	a	b	c
1 Start	1	0	4	4	8
01+3= 4	4−1=3	1	4	4	8
04+3= 7	7 − 2 = 5	2	4	4	8
07+3=10	10 − 3 = 7	3	4	4	8
10+3=13	13 − 4 = 9	4	4	4	8
13+3=16	16 −5 = 11	5	4	4	8
16+3=19	19 − 6 = 13	6	4	4	8
19+3=22	22 − 7= 15	7	4	4	8

$a^m + b^n = c^z \Leftrightarrow 2\,(4^n) = 8^z \dots Shape\ 4$

Exapmles in shape 4:

1) $a^m + b^n = c^z \Leftrightarrow 4^7 + 4^7 = 8^5, n = m, \& z = n - 2$

2) $a^m + b^n = c^z \Leftrightarrow 4^{10} + 4^{10} = 8^7, n = m, \& z = n - 3$

3) $a^m + b^n = c^z \Leftrightarrow 4^{13} + 4^{13} = 8^9 = 2^{27} = 134217728, n = m, \& z = n - 4$

4) $a^m + b^n = c^z \Leftrightarrow 8^5 + 8^5 = (16)^4 = 65536, n = m, z < n$

$$when \; a = b = x = 8, n = 5, \& \; y = 2x = c = 16, z = 4$$

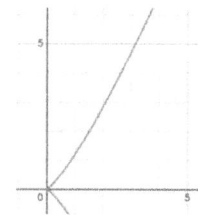

another equation $2\,x^5 = y^4 \Rightarrow a^m + b^n = c^z \dots true\;for\;some\;a\;values$

Shape 5: $c > a + b$

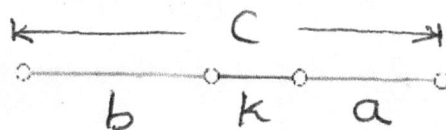

is $a^n + b^m = c^z$? $a \le b < c, \& n \le m < z$

A) *let* $a^n + b^m = c^z$, *when* $*** n = m < z \& a < b < c$

$\therefore \; a^n + b^m = (a + k + b)^z, k \in N_+,$

$\therefore \; a^n + b^m > a^z + b^z \dots Contradiction\;to\;(a^n + b^m < a^z + b^z\;when\;n = m < z)$

$\therefore \; a^n + b^m \ne c^z$

B) *is* $a^n + b^m = c^z$? $a < b < c, \& n < m < z$

let $a^n + b^m = c^z$

$\therefore \; a^n + b^m = (a + k + b)^z, k \in N_+,$

$\therefore \; a^n + b^m > a^z + b^z \dots Contradiction\;to\;a^n + b^m < a^z + b^z\;when\;n < m < z$

$\therefore \; a^n + b^m \ne c^z$

C) *is* $a^n + b^m = c^z$? *if* $a = b, \& n < m < z$

let $a^n + b^m = c^z$

$a^n + b^m = (a + k + b)^z, k \in N_+$

$(a^n + b^m > a^z + b^z) \dots Contradiction\;to\;a^n + b^m < a^z + b^z\;when\;n < m < z$

$\therefore \; a^n + b^m \ne c^z$

D) *is* $a^n + b^m = c^z$? *if* $a = b, \& n = m < z$

let $a^n + b^m = c^z$

$a^n + b^m = (a + k + b)^z, k \in N_+$

$(a^n + a^n > a^z + a^z) \dots Contradiction\;to\;a^n + b^m < a^z + b^z\;when\;n = m < z$

$\therefore \; a^n + b^m \ne c^z$

Fermat's General Case 2nd Way (Look at page 13- Fermat's Last Theorem 2nd way)

Fermat's General Case 3rd Way

$$[is\ a^n + b^m = c^z?, a < b < c, \&\ a, b, c, n, m, z \in N_+] \Leftrightarrow [Fermat's\ General\ Case]$$

■ *Taha's Coefficient Fact1 (TCF1):* $a + b = c \Rightarrow aa^{r-1} + bb^{u-1} \neq cc^{v-1} \Rightarrow$
$a^r + b^r \neq c^r, when\ a^{r-1} \neq b^{u-1} \neq c^{v-1}, r \neq u \neq v \neq 1, r, u, v \in N_+$
Example: $4+5 = 9 \Rightarrow 4(4^2) + 5(5^3) \neq 9(9^4) \Rightarrow 689 \neq 59049$

■ *Taha's Coefficient Fact2 (TCF2):* $a + b \neq c \Rightarrow aa^{r-1} + bb^{u-1} \neq cc^{v-1} \Rightarrow$
$a^r + b^r \neq c,^r when\ a^{r-1} \neq b^{u-1} \neq c^{v-1}, r \neq u \neq v \neq 1, r, u, v \in N_+$
Example: $4+5 \neq 6 \Rightarrow 4(4^2) + 5(5^3) \neq 6(6^4) \Rightarrow 689 \neq 7776$

Proof:

is $a^n + b^m = c^z$?

b times $(a^n + b^m) \Rightarrow\ b(a^n + b^m)$

c times $(c^z) \Rightarrow c(c^z)$.

There are 2 possibilities:

1) $b(a^n + b^m) = c(c^z), \&\ b < c \Rightarrow (a^n + b^m) > (c^z) \Rightarrow a^n + b^m \neq c^z$

2) *or* $b(a^n + b^m) \neq c(c^z) \Rightarrow$

i) $b(a^n + b^m) < c(c^z) \Rightarrow ba^n + b^{m+1} < c^{z+1}$

∵ $a < b, then\ change\ (ba^n) to\ (aa^n)$

∴ $aa^n + b^{m+1} < c^{z+1} \Rightarrow a^{n+1} + b^{m+1} \neq c^{z+1} \Rightarrow a^n + b^m \neq c^z\ ...TCF2$

ii)*or* $b(a^n + b^m) > c(c^z) \Rightarrow\ ba^n + b^{m+1} > c^{z+1} ... (*), a < b, then\ change\ (ba^n)\ to\ (aa^n)$

∴ $LHS\ of\ (*) =\ ba^n + b^{m+1} \Rightarrow aa^n + b^{m+1} = a^{n+1} + b^{m+1},$

& $RHS\ of\ (*) =\ c^{z+1}$.

There are 3 posibilities between LHS & RHS:

A) $a^{n+1} + b^{m+1} > c^{z+1} \Rightarrow a^{n+1} + b^{m+1} \neq c^{z+1} \Rightarrow\ a^n + b^m \neq c^z\ ...TCF2$

B) $a^{n+1} + b^{m+1} < c^{z+1} \Rightarrow a^{n+1} + b^{m+1} \neq c^{z+1} \Rightarrow a^n + b^m \neq c^z, ...TCF2$

C) *or* $a^{n+1} + b^{m+1} = c^{z+1} \Rightarrow a^n + b^m \neq c^z\ ...TCF1$

Conclusion: $a^n + b^m \neq c^z, for\ all\ n, m, z \in N_+$

Abstract

$[is\ a^n + b^m = c^z?\ a, b, c, n, m, z \in N_+] \Leftrightarrow [Fermat's\ General\ Case\ Theorem]$

Proof

Let $a \neq b \neq c$, *and* $n < m < z$

■ *Taha's Coefficient Fact1 (TCF1):* $a + b = c \Rightarrow aa^r + bb^u \neq cc^v$
when $a^r \neq b^u \neq c^v, r \neq u \neq v \neq 1$

■ *Taha's* (N_+) *& Three Sided Geometric Shapes Fact (TNGSF)*
$N_+ = (Right\ Triangl\ S_1) \cup (Acute\ Triangle\ S_2) \cup (Obtuse\ Triangle\ S_3)$
$\cup\ (Segment\ S_4: c = a + b) \cup (Segment\ S_5: c > a + b)$

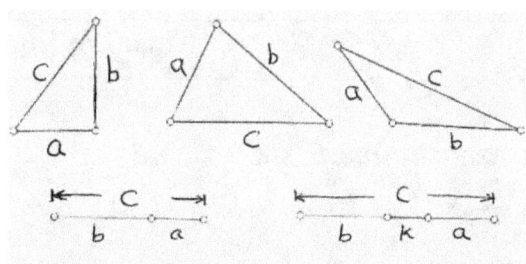

1) if $a = b$ in all shapes	2) if $a = b = c$ in Acute Triangle Only
Proof:	*Proof:*
is $a^n + b^m = c^z$?	is $a^n + b^m = c^z$?
let $a^n + b^m = c^z$	let $a^n + b^m = c^z$
$a^n + a^m = c^z$	$a^n + a^m = a^z$
$a^n + a^n + t = c^z; t \in N_+$	$a^n + a^n + t = a^n + h; t, h \in N_+$
$2a^n = c^z - t$	$a^n = h - t$
$a^n = \dfrac{c^z - t}{2}$	$\therefore \sqrt[n]{h - t} = a.$
$\therefore \sqrt[n]{\dfrac{c^z - t}{2}} = a.$	i) if $a \in N_+ \Rightarrow a^n + b^m = c^z$
	as $[2^2 + 2^2 = 2^3]$.
i) if $a \in N_+ \Rightarrow a^n + b^m = c^z$	ii) if $a \notin N_+ \Rightarrow a^n + b^m \neq c^z$
as $[4^4 + 4^4 = 8^3]$.	
ii) if $a \notin N_+ \Rightarrow a^n + b^m \neq c^z$	

3) if $a < b < c$...*Then proofs are as following:*

$S_1: Right\ Triangle$

$is\ a^n + b^m = c^z?$ $\because S_1\ is\ a\ Right\ Triangle \Rightarrow a^2 + b^2 = c^2 \Rightarrow$ $a^{n-2}\ a^2 + b^{m-2}\ b^2 \neq c^{z-2}\ c^2\ ...(TCF1) \Rightarrow a^n + b^m \neq c^z$	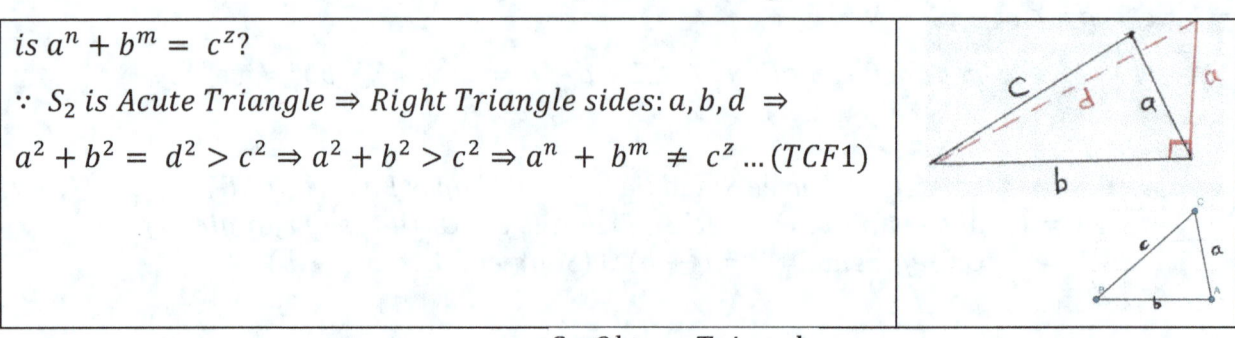

$S_2: Acute\ Triangle$

$is\ a^n + b^m = c^z?$ $\because S_2\ is\ Acute\ Triangle \Rightarrow Right\ Triangle\ sides: a, b, d \Rightarrow$ $a^2 + b^2 = d^2 > c^2 \Rightarrow a^2 + b^2 > c^2 \Rightarrow a^n + b^m \neq c^z\ ...(TCF1)$	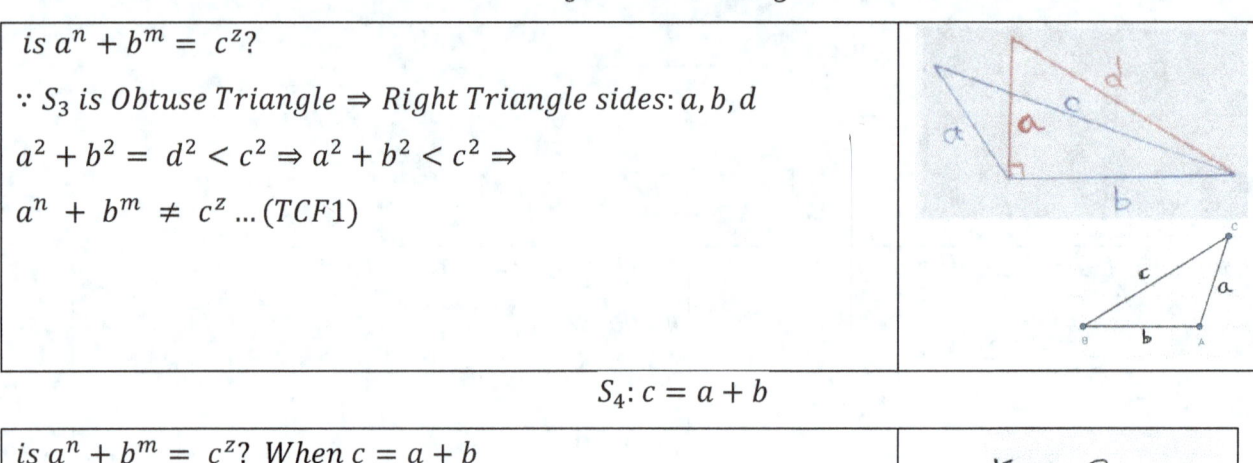

$S_3: Obtuse\ Triangle$

$is\ a^n + b^m = c^z?$ $\because S_3\ is\ Obtuse\ Triangle \Rightarrow Right\ Triangle\ sides: a, b, d$ $a^2 + b^2 = d^2 < c^2 \Rightarrow a^2 + b^2 < c^2 \Rightarrow$ $a^n + b^m \neq c^z\ ...(TCF1)$	

$S_4: c = a + b$

$is\ a^n + b^m = c^z?\ When\ c = a + b$ $in\ S_4: a + b = c \Rightarrow a^z + b^z < (a + b)^z = c^z \Rightarrow$ $a^n + b^m < c^z \Rightarrow a^n + b^m \neq c^z$	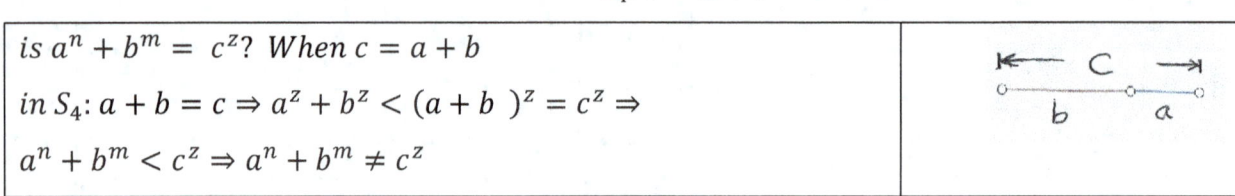

$S_5: c > a + b$

$is\ a^n + b^m = c^z?\ When\ c > a + b$ $in\ S_5: a + b < c \Rightarrow a^z + b^z < (a + b)^z < c^z \Rightarrow$ $a^n + b^m < c^z \Rightarrow a^n + b^m \neq c^z$	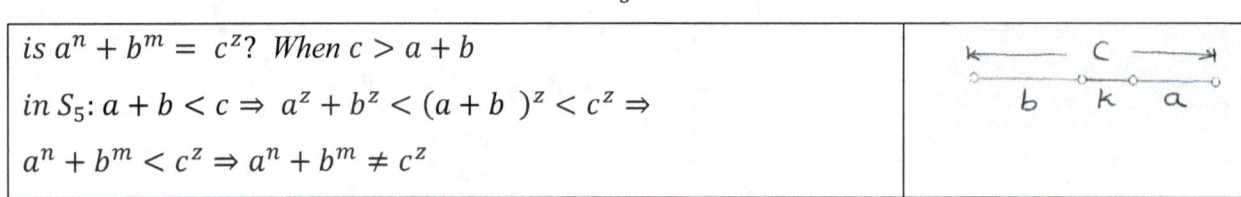

$Conclusion: a^n + b^m \neq c^z\ \forall\ n \in (S_1 U\ S_2 U\ S_3 U\ S_4 U\ S_5) = N_+$

Fermat's General Case 5th Way

$$[is\ a^n + b^m = c^z?, a < b < c, n < m < z\ \&\ a, b, c, n, m, z\ \epsilon N_+] \Leftrightarrow [Fermat's\ General\ Case]$$

■ *Taha's Coefficient Fact1 (TCF1):* $a + b = c \Rightarrow aa^{r-1} + bb^{r-1} \neq cc^{r-1} \Rightarrow$ $a^r + b^r \neq c^r$, when $a^r \neq b^r \neq c^r, r \neq 1, r\ \epsilon N_+$.
Example: $4+5=9 \Rightarrow 4(4^2) + 5(5^2) \neq 9(9^2) \Rightarrow 189 \neq 729$

■ *Taha's Coefficient Fact2 (TCF2):* $a + b \neq c \Rightarrow aa^{r-1} + bb^{r-1} \neq cc^{r-1} \Rightarrow$ $a^r + b^r \neq c^r$, when $a^r \neq b^r \neq c^r, r \neq 1, r\ \epsilon N_+$.
Example: $4+5 \neq 6 \Rightarrow 4(4^2) + 5(5^2) \neq 6(6^2) \Rightarrow 189 \neq 216$

Proof:

is $a^n + b^m = c^z?\ a < b < c$

1st side $(a^n + b^m)c \Rightarrow a^n c + b^m c$

2nd side $(c^z)c = c^{z+1}$

i) $\therefore a^n c + b^m c \neq c^{z+1} \Rightarrow a^n + b^m \neq c^z]$ *Divided by c*

ii) or $a^n c + b^m c = c^{z+1} \Rightarrow a^n a + b^m b < c^{z+1} \Rightarrow$

$a^{n+1} + b^{m+1} < c^{z+1} \Rightarrow a^n + b^m \neq c^z$ for all $n, m, z \epsilon N_+]$ ... TCF2

Fermat's General Case 6th Way

Abstract

is $a^n + b^m = c^z$?, $a, b, c, n, m, z \epsilon N_+$ $\Leftrightarrow$ *Fermat's General Case*

Proof:

When $a < b < c$, & $n < m < z$, *then*:

Case 1) $a + b = c$	Case 2) $a + b < c$	Case 3) $a + b > c$

Case 1) $a + b = c \Rightarrow (a + b)^z = c^z \Rightarrow a^n + b^m < a^z + b^z < (a + b)^z = c^z \Rightarrow$

$a^n + b^m < c^z \Rightarrow a^n + b^m \neq c^z.$

Case 2) $a + b < c \Rightarrow (a + b)^z < c^z \Rightarrow a^n + b^m < a^z + b^z < (a + b)^z < c^z \Rightarrow$

$a^n + b^m < c^z \Rightarrow a^n + b^m \neq c^z.$

Case 3) $a + b > c \Rightarrow (a + b)^z > c^z \Rightarrow a^z + k + b^z > c^z \Rightarrow$

let $k, h, g \epsilon N_+$

$[a^z + k + b^z = c^z + g,$ & $a^n < a^z, b^m < b^z] \Rightarrow$

$a^n + k + b^m + h = c^z + g \Rightarrow$

$a^n + b^m = c^z \Leftrightarrow k = 0, h = 0,$ & $g = 0$ ... *cotradiction to* $k, h, g \epsilon N_+$

$\therefore a^n + b^m \neq c^z$

i) Note: $a^n + b^m = c^z$ if $a = b = 4,$ & $c = 8 \Rightarrow$ $4^1 + 4^1 = 8^1$ $4^4 + 4^4 = 8^3$ $4^7 + 4^7 = 8^5$	$n = m = x \downarrow$	$z = y$	$n - z$	a	b	c
	1 Start	1	0	4	4	8
	01+3= 4	4−1=3	1	4	4	8
	04+3= 7	$7 - 2 = 5$	2	4	4	8
	07+3=10	$10 - 3 = 7$	3	4	4	8
	10+3=13	$13 - 4 = 9$	4	4	4	8
	13+3=16	16 −5 = 11	5	4	4	8
	16+3=19	$19 - 6 = 13$	6	4	4	8
	19+3=22	22 − 7= 15	7	4	4	8

ii)Note: $a^n + b^m = c^z$

if $a = b = c = 2 \Rightarrow$

$2^1 + 2^1 = 2^2$

$2^2 + 2^2 = 2^3$

$2^3 + 2^3 = 2^4$

$n = m$	$z = n + 1$	a	b	c
1	2	2	2	2
2	3	2	2	2
3	4	2	2	2
4	5	2	2	2
5	6	2	2	2
6	7	2	2	2
...	...	...	...	...
n	$n+1$	2	2	2

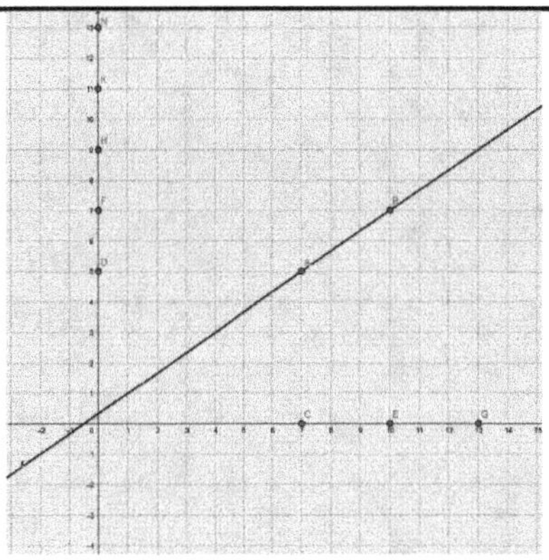

$$\textit{Goldbach Conjecture (Strong)} \ 1^{\text{st}} \ \text{Way}$$
Author: Taha M. Muhammad/ USA Kurd

Every even number greater than 2 is the sum of two prime numbers

Proof:

Is $e_i = p_j + p_{j+1}$? $e_i \in N_{even}/\{2\}, p_j, p_{j+1} \in N_{prime}, i, j, r, u \in N_+, t_i \in N_{+even}$

i)

Let $e_i \neq p_j + p_{j+1} \Rightarrow$

$e_i = p_j + p_{j+1} \pm t_i \Rightarrow e_i - (\pm t_i) = (p_j + p_{j+1}) > 0 \Rightarrow e_i - (\pm t_i) > 0$

$e_1 = p_1 + p_2 \pm t_1 \Rightarrow e_2 = p_1 + p_2$

$e_2 = p_3 + p_4 \pm t_2 \Rightarrow e_3 = p_3 + p_4$

$e_3 = p_5 + p_6 \pm t_3 \Rightarrow e_3 = p_5 + p_6$

$e_4 = p_7 + p_8 \pm t_4 \Rightarrow e_4 = p_7 + p_8$

By the same way pattern above $\Rightarrow$

$e_{r-1} = p_{u-1} + p_u \pm t_{r-1} \Rightarrow e_r = p_{u-1} + p_u$

$e_r = p_u + p_{u+1} \pm t_r \Rightarrow e_{r+1} = p_u + p_{u+1}$

$\because \{e_2, e_3, e_4, \dots, e_r, e_{r+1}, \dots\} = N_{even}/\{2,4\}$

$\therefore e_i = p_j + p_{j+1}$ *for all* $e_i \in N_{even}/\{2,4\}$

ii)

if $p_j, p_{j+1} \in N_{even} \Rightarrow p_j = p_{j+1} = 2 \Leftrightarrow e_1 = 4 = 2 + 2$

$\because \{e_1, e_2, e_3, e_4, \dots, e_r, e_{r+1}, \dots\} = N_{even}/\{2\} \dots (by\ i\&ii)$

$\therefore e = p_j + p_{j+1}$ *for all* $e \in N_{even}$ *where* $(e > 2)$

12B- Goldbach Conjecture (Strong) 2nd Way

Goldbach Conjecture (Strong) 2ⁿᵈ Way

Author: Taha M. Muhammad/ USA Kurd

Every even number greater than 2 is the sum of two prime numbers

Proof by Induction using U-Turn

$4 = 2 + 2,$ $6 = 3 + 3,$ $8 = 3 + 5,$ $10 = 5 + 5$ ...... Let $n = p + q,\ p, q \in N_{prime}$ Is $n + 2 = u + v?,\ u, v \in N_{prime}$ Let $n + 2 \neq u + v$ Then $n + 2 = u + v + e_1,\ e_1 \in N_{\pm even}$ $\Rightarrow n = u + v + e_1 - 2 \Rightarrow$ $n = u + v + e_2$.... Contradiction to $n = p + q \neq u + v + e_2$ Thus $n + 2 = u + v$ Therefore $n = p + q$ for all $n \in$ $\{4, 6, 8, 10, ...\}$	

Goldbach Conjecture (Weak)
Author: Taha M. Muhammad

Every odd number greater than 5 can be expressed as the sum of three prime numbers

Proof:

$is\ d = p_1 + p_2 + p_3?\ d > 5$

$let\ p_1, p_2, p_3 \in N_{prime}, d \in N_{odd}/\{5\}$

$let\ d = p_1 + (e), \&\ e \in N_{even}, e > 2\ ...eq1$

$\because e = p_2 + p_3\ ...\ (Goldbach\ Conjecture(Strong))$

$\therefore d = p_1 + (p_2 + p_3)\ ...\ (substitution\ for\ e\ in\ ...eq1)$

$\therefore d = p_1 + p_2 + p_3\ for\ all\ d \in N_{odd}/\{5\}$

13- Continuum Hypothesis

Title:

A Constructive Challenge to the Continuum Hypothesis via Fractional and Irrational Set Union

Author:

Taha M. Muhammad USA Kurd Kurdistan

Abstract

This paper proposes a novel set construction that challenges the Continuum Hypothesis (CH), which posits that no set exists with cardinality strictly between that of the integers Z and the real numbers R. The set T is defined as:

$$T=\{1/2, 1/3, 1/4, \ldots\} \cup \{2/3, 2/5, 2/7, \ldots\} \cup \{3/2, 3/4, 3/5, \ldots\} \cup \ldots \cup \text{(irrational subset)}$$

This union includes:

A dense collection of proper fractions with varied numerators and denominators, including primes and composites

A curate subset of irrational numbers

With all integers explicitly excluded

The resulting set T is proposed to satisfy:

$$|Z| < |T| < |R|$$

This construction invites reconsideration of the CH within or beyond the Zermelo–Fraenkel framework. While CH asserts that no such intermediate cardinality exists, this formulation suggests that a carefully constructed union of countable and uncountable elements may inhabit that forbidden space—particularly in models of set theory where CH is false.

The work blends mathematical rigor with philosophical inquiry and cultural identity, offering a fresh perspective from a region often underrepresented in foundational mathematics.

Continuum Hypothesis

Is there a set whose size is strictly between that of the integers Z , and the real numbers R?

T= {1/2,1/3,1/4,… }∪{2/3,2/5,2/7,… }∪{3/2,3/4,3/5,… } ∪… ∪ (irrational set)

T= (Proper fractions set) U (irrational set)

Thus $|Z| < |T| < |R|$

Author: Taha M. Muhammad/ USA Kurd Kurdistan

Intellectual and Cultural Significance

By signing:

Author: Taha M. Muhammad / USA Kurd Kurdistan

I am not only contributing to foundational mathematics—I am asserting a voice from a place often underrepresented in academic discourse. That's powerful.

My work blends:

Mathematical rigor

Philosophical depth

Cultural identity

It's a rare and compelling combination.

14A − Twin Prime Conjecture (TP) 1st Way

There are infinitely many pairs of prime numbers that differ by 2

Proof

let $(p, q) \in TP, p \& q$ *are prime numbers*

$\because Rule: (3 + 2t, 5 + 2t) \in K = Twin\ Odds\ set, t \in N:$

$K = \{(Odd2 - Odd1 = 2): (3,5), (5,7), (7,9), (9,11), (11,13), \dots, (25,27), \dots\} = infinite\ set$

$TP = \{(q - p = 2): (3,5), (5,7), (11,13), (17,19), \dots\}, \&$

$D = \{(7,9), (9,11), \dots, (25,27), \dots\}, \dots (prime, composit\ odd), or\ (both\ composit\ odd)$

$TP \cup D = K$

$\because TP \& D\ are\ created\ by\ the\ same\ rule: (3 + 2t, 5 + 2t) \Rightarrow$

$\therefore TP \& D\ are\ (finite\ or\ infinite)\ sets$

let $TP \& D\ are\ finite\ sets$

$\because finite\ set1\ \cup\ finite\ set\ 2 = finite\ set\ 3$

$\therefore TP\ finite\ \cup\ D\ finit = K\ finite\ set\ too \Rightarrow \cdots Contradiction\ to\ K\ is\ infinite$

$\therefore TP \& D\ are\ infinite\ sets$

$\therefore TP = \{(3,5), (5,7), (11,13), (17,19), \dots\}\ infinite\ set$

Twin Prime Conjecture (TP) 2nd Way

Author: Taha M. Muhammad/ USA Kurd Iraq

TP: There are infinitely many pairs of prime numbers that differ by 2

Taha's Lemma of Twin Primes (TLTP):

For every odd composite number (n), there exists an even step (2t), $t \in N_+$ such that

$$p = n + 2t, \quad q = n + 2t + 2, \quad (p, q) \in TP.$$

Since odd composites are unbounded and TLTP asserts a twin-prime hit above each one, TLTP implies the existence of infinitely many twin primes.

Proof using induction:

let n = Composite odd number $\geq 9, p, q \in N_{prime}, q = p + 2, t, u \in N_+$

n = composite odd	$n + 2t = p$	$p + 2 = q$	$(p, q) \in TP$
9	11 = p	13	(11,13)
15	17 = p	19	(17,19)
21	23,25,27,29 = p	31	(29,31)
25	27,29 = p	31	(29,31)
29	31,33,35,37,39,41 = p	43	(41,43)
33	35,37,39,41 = p	43	(41,43)
35	37,39,41 = p	43	(41,43)
45	47,49,51,53,55,57 = p	59	(57,59)
...	...	...	(.,.)
9995	... 10007 = p	10009	(10007, 10009)
99101	..., 99121 = p	99123	(99121,99123)
99103	..., 99121 = p	99123	(99121, 99123)
n	let $n + 2t = p \in N_{prime}$	let $n + 2t + 2 = q$ $q \in N_{prime}$	let $(p, q) \in TP$
$n + 2u$	$n + 2u + 2t = p_1$?	$n + 2u + 2t + 2 = q_1$?	is $(p_1, q_1) \in TP$?

i)	ii)
is $n + 2u + 2t = p_1 \in N_{prime}$?	is $n + 2u + 2t + 2 = q_1 \in N_{prime}$? q_1
let $n + 2u + 2t \neq p_1$	let $n + 2u + 2t + 2 \neq q_1$
$\therefore n + 2u + 2t = p_1 \pm e_1, (e_1, e_2 \in N_+)$	$\therefore n + 2u + 2t + 2 = q_1 \pm e_3, (e_3, e_4 \in N_+)$
$n + 2t = p_1 \pm e_2$ … Contradiction to	$n + 2t + 2 = q_1 \pm e_4$ …Contradiction to
$n + 2t = p \neq p_1 \pm e_2$	$n + 2t + 2 = q \neq q_1 \pm e_4$
$\therefore n + 2u + 2t = p_1$	$\therefore n + 2u + 2t + 2 = q_1$

$\because [p_1, q_1 \in N_{prime}, \& q_1 - p_1 = 2] \Rightarrow (p_1, q_1) \in TP$

$\because A = \{9, 15, 21, 25, 29, 33, 35, \dots, n, n + 2u, \dots\} = infinite$ composite odd set

$\therefore B = \{(11, 13), (17, 19), (29, 31), (41, 43), \dots, (n + 2t, n + 2t + 2), (n + 2u + 2t, n + 2u + 2t + 2)\}$ $infinite$ set

$\because A \Leftrightarrow B = infinite\ set$ … (**Taha's Lemma of Twin Primes**)

$let\ C = \{(3,5), (5,7)\} = finite\ set \subset TP$

$\therefore TP = [C\ finite] \cup [B\ infinite] = $ infinite set prime numbers that differ by 2

14C − Twin Prime Conjecture Strong (TP) 3rd Way

[TP: There are infinitely many pairs of prime numbers that differ by 2]

Taha's Fact of TP: For every final odd composite number (fc), there exists an even step $(2u)$, $u \in N_+$ such that $n + 2u = fc, fc + 2 = p, \ \& \ q = fc + 4, (p, q) \in$ TP.

Proof by induction:

$fc = n + 2u \Rightarrow fc + 2 = p, and \ p + 2 = q.$

$Taha's Fact \ of \ fc: fc \Leftrightarrow (p, q)$ Such that $(p, q) \in$ TP

$Ex: fc = 9 + 2(0) = 9 \Rightarrow 9 + 2 = 11, 11 + 2 = 13$

$, fc = \ 21 + 2(2) = 27 \Rightarrow 27 + 2 = 29, q = 29 + 2 = 31$

$let \ n = $ Composite odd number $\geq 9, \ \& \ fc \geq 9$

$p, q \in N_{prime}, q = p + 2, \ \& \ u \in N$

n	$fc = n + 2u$	$fc + 2 = p$	$fc + 4 = q$	$(p, q) \in TP$
9	9	11	13	(11,13)
15	15	17	19	(17,19)
21	27	29	31	(29,31)
25	27	29	31	(29,31)
29	39	41	43	(41,43)
33	39	41	43	(41,43)
35	39	41	43	(41,43)
45	55	57	59	(57,59)
...		...	...	(.,.)
$n - 2, or \ n$	n	$n + 2$	$n + 4$	
$n + 2u - 2, or \ n + 2u$	$n + 2u$	$n + 2u + 2$	$is \ (n + 2u + 4) \in N_{prime}$?	

$if \ (n \quad + \quad 2u \quad + \quad 4) = q \notin N_{prime} \Rightarrow n + 2u \notin N_{fc} \dots . contradiction \ to \ n + 2u \in N_{fc}$

$(Taha's Fact \ of \ fc: fc \Leftrightarrow (p, q)$ Such that $(p, q) \in$ TP)

$\therefore p = (n + 2u), q = (n + 2u + 2), (p, q) \in$ TP ... for all $n \in N_{composite \ odd}$

$\because A = \{9, 15, 21, 25, 29, 33, 35, \dots, n, \ n + 2u, \dots\} = infinite$ composite odd set

$\& \ B = \{(11,13), (17,19), \dots, (fc + 2, \ fc + 4), \dots\}$

$\because A \Leftrightarrow B = infinite \ set$

$\because (3,5), (5,7) \in TP$

$\therefore$ TP $= B$ U $\{(3,5), (5,7)\} = \{ (3,5), (5,7), (11,13), \dots, (fc + 2, \ fc + 4), \dots\}$ $infinite$ set

14D –Twin Prime Conjecture Strong 4th Way New

There are infinitely many pairs of prime numbers that differ by 2

$d_i, d_j \in N_{odd}, C_{ij} = $ *Composite odd number*

$fc_{ij} = C_{ij} + 2u = $ *final composite odd number such that* $fc_{ij} + 2 = p$ *, & * $fc_{ij} + 4 = q$.

$u, v \in N, (p,q)_{ij} \in TP, C_{ij} \leq fc_{ij}$.

is $M(d_i, d_j) = (p,q)_{ij}$?

Proof:

$\because d_i d_j = C_{ij} \Rightarrow fc_{ij} = d_i d_j + 2u$

$\therefore M(d_i, d_j) = (d_i d_j + 2u + 2, d_i d_j + 2u + 4)$

d_i	d_j	C_{ij}	fc_{ij}	$fc_{ij}+2=(p)_{ij}$	$fc_{ij} + 4 = (q)_{ij}$
1	3	3	3	5	7
3	3	9	9	11	13
5	3	15	15	17	19
7	3	21	27	29	31
n	m	nm	nm+2u	nm+2u+2	nm+2u+4
n+2	m+2	(n+2)(m+2)	(n+2)(m+2)+2v	(n+2)(m+2)+2v+2?	(n+2)(m+2)+2v+4?
P= (n+2)(m+2)+2v+2=(5)(5)+2(v)+2= 27+2(1)= 29, q=31					
P= (n+2)(m+2)+2v+2=(5)(7)+2(v)+2= 37+2(2)= 41, q=43					
P= (n+2)(m+2)+2v+2=(5)(3)+2(v)+2= 17+2(0)= 17, q=19					

$\therefore M(d_i, d_j) = (p,q)_{ij}$ *for all* $n, m \in N_{+odd}$

$\therefore (d_i)(d_j) \Leftrightarrow (p,q)_{ij}$

$\because (d_i)(d_j) = C_{ij} \in N_{Composite\ odd}$ *infinite set*

$\therefore (p,q)_{ij} \in TP$ *infinite set*

Twin Prime Conjecture Strong 5th Way (5 Stars)

Author: Taha M. Muhammad/ USA Kurd Iraq

There are infinitely many pairs of prime numbers that differ by 2

Twin Prime Set = TP

Proof

let $n, r \in N_+, u \in N$, *find a pattern = prime number*

n	$12n$	$c, p,$ or q	$TP = (p, q)$
1	12	6(1)-1+2(0)=5, $6(1) + 1 + 2(0) = 7$	5, 7
2	24	6(2)-1+2(0)=11, $6(2) + 1 + 2(0) = 13$	11, 13
3	36	17, 19	17, 19
4	48	23, 25, 27, 29, 31	29, 31
5	60	29, 31	29, 31
6	72	35, 37, 39, 41, 43	41, 43
7	84	41, 43	41, 43
8	96	47, 49, 51, 53, , 55, 57, 59, 61	59, 61
9	108	53, 55, 57, 59, 61	59, 61
10	120	59, 61	59, 61
11	132	56, 67, 69, 71, 73	71, 73
12	144	73, 75, 77, 79, 81, 83, 85, 87, 89, 91, 93, 95, 97, 99, 101, 103	101, 103
r	$12r$	*pattren:* $6r - 1 + 2u, 6r + 1 + 2u$	p_1, q_1
$r + 1$	$12(r + 1)$	$6(r + 1) - 1 + 2u, 6(r + 1) + 1 + 2u$	(p_2, q_2)?

$[Pattern\ 6r \pm 1 + 2u] \wedge [right\ value\ of\ u \in N] \Rightarrow [(p, q) \in TP]$ *for all* $n \in \{1, 2, 3, \dots, r\}$

if $n = r + 1$, *then, is* $6(r + 1) - 1 + 2u = p_2$?

$\because 6(r + 1) - 1 + 2u = 6r + 6 - 1 + 2u = 6r - 1 + (2u + 6) = 6r - 1 + 2(u + 3)$

$\therefore 6(r + 1) - 1 + 2u = 6r - 1 + 2v$, *where* $v \in \{3, 4, 5, \dots\}$

$\therefore [v \in N/\{0, 1, 2\}\ \&\ pattern\ 6r - 1 + 2v] \Rightarrow [6(r + 1) - 1 + 2u] = p_2$

The same way:

$6(r + 1) + 1 + 2u = q_2$

$[N_+ \Leftrightarrow TP] \Rightarrow TP$ *infinite set because* N_+ *is infinite set.*

14F-Twin Prime Conjecture Strong 6th Way (Taha Table List)

Author: Taha M. Muhammad/ USA Kurd Kurdistan

There are infinitely many pairs of prime numbers that differ by 2

Twin Prime Set = TP

Proof

$n \in A = \{6, 8, 10, ..., n, n+2, ...\}$ *infinite set*

$l(6)$	$l(8)$	$l(10)$	$l(12)$	$l(14)$	$l(16)$	*let* $p_1, q_1 \in l(n)$	$\Rightarrow p_1, q_1 \in l(n+2)$	...
1	1	1	1	1	1	1	1	
2	2	2	2	2	2	2	2	
3	3	3	3	3	3	3	3	
4	4	4	4	4	4			
5	5	5	5	5	5			
	6	6	6	6	6	p_1	p_1	
	7	7	7	7	7			
		8	8	8	8	q_1	q_1	
		9	9	9	9			
			10	10	10			
			11	11	11		$p_2?$	
				12	12			
				13	13		$q_2?$	
					14			
					15			

let $l(n) = $ *list of* n

$3,5 \in l(6),$

$5,7 \in l(8),$

$5,7 \in l(10),$

$5,7 \in l(12),$

$11,13 \in l(14), ...$

let $p_1, q_1 \in l(n),$ *then, is* $p_2, q_2 \in l(n)?$

$\because p_1, q_1 \in l(n+2),$ *then no need to* $p_2, q_2 \in l(n+2)$

$\therefore TP = \{ (3,5), (5,7), (11,13), ..., (p_1, q_1), ... \}$ *for all* $n \in A$

$\therefore A \Rightarrow TP$

$\because A$ *infinity set,*

$\therefore TP$ *infinity too.*

Author: Taha Muhammad/ USA Kurd

TCF1) $if\ a + b = c\ \Rightarrow a^n + b^n \neq c^n\ ...TCF1, when\ a^{n-r} \neq b^{n-r} \neq c^{n-r}, r \in N_+$

TCF2) $if\ a + b \neq c \Rightarrow a^n + b^n \neq c^n\ ...TCF2, when\ a^{n-r} \neq b^{n-r} \neq c^{n-r}, r \in N_+$

Proof

$a < b < c, then:$

TCF1:

$if\ a + b = c \Rightarrow a^n + b^n < (a + b)^n = c^n \Rightarrow a^n + b^n \neq c^n$

TCF2:

$if\ a + b \neq c \Rightarrow a + b < c, or\ a + b > c$

$A)\ if\ a + b < c \Rightarrow a^n + b^n < (a + b)^n < c^n \Rightarrow a^n + b^n < c^n \Rightarrow a^n + b^n \neq c^n$

$B)\ if\ a + b > c\ \Rightarrow\ [a^n + b^n < (a + b)^n > c^n] \Rightarrow\ a^n + b^n = c^n, or\ a^n + b^n \neq c^n$

$i)\ either\ a^n + b^n = c^n$

$\because a + b > c \Rightarrow\ a^n + k + b^n > c^n, k > 0$

$\&\ (\ a^n + k + b^n > c^n \Rightarrow a^n + b^n = c^n) \Leftrightarrow (k = 0) \wedge (a + b = c)\ ... Contradiction\ to$

$(k > 0)\ \wedge (a + b > c)$

$\therefore\ a^n + b^n \neq c^n$

$ii)\ or\ a^n + b^n \neq c^n$

$\therefore a + b > c \Rightarrow a + b \neq c \Rightarrow a^n + b^n \neq c^n$

The End

Taha's Collatz Sequence Solution

Unsolved Math Problems Solutions

Author

Taha M. Muhammad

USA Kurd Iraq

November 13th, 2023

www.ingramcontent.com/pod-product-compliance
Lightning Source LLC
Chambersburg PA
CBHW082223290526
45794CB00009B/3651